Energy Security, Equality, and Justice

This book applies concepts from ethics, justice, and political philosophy to five sets of contemporary energy problems cutting across time, economics, politics, geography, and technology.

In doing so, the authors derive two key energy justice principles from modern theories of distributive justice, procedural justice, and cosmopolitan justice. The *prohibitive principle* states that "Energy systems must be designed and constructed in such a way that they do not unduly interfere with the ability of people to acquire those basic goods to which they are justly entitled." The *affirmative principle* states that "If any of the basic goods to which people are justly entitled can only be secured by means of energy services, then in that case there is also a derivative entitlement to the energy services." In laying out and employing these principles, the book details a long list of current energy injustices ranging from human rights abuses and energy-related civil conflict, to energy poverty and pervasive and growing negative externalities.

The book illustrates the significance of energy justice by combining the most up-to-date data on global energy security and climate change, including case studies and examples from the electricity supply, transport, and heating and cooking sectors, with appraisals based on centuries of thought about the meaning of justice in social decisions.

Benjamin K. Sovacool is Director of the Danish Center for Energy Technology at AU-Herning and a Professor of Business and Social Sciences at Aarhus University in Denmark. He is also Associate Professor of Law at Vermont Law School and Director of the Energy Security and Justice Program at their Institute for Energy and the Environment.

Roman V. Sidortsov is currently a Senior Global Energy Fellow at the Institute for Energy and the Environment of Vermont Law School, as well as a Doctoral Researcher at the Scott Polar Research Institute, Cambridge University, United Kingdom.

Benjamin R. Jones is currently a Senior Global Energy Fellow at the Institute for Energy and the I   as a Doctoral Candidate in the Fa   nada.

"Many aspects of growing worldwide energy consumption have sparked debate and discussion, but very little attention has been paid to the social and ethical dimensions of this issue – despite the fact that these aspects are certain to a play an increasingly critical role as doubts arise over the adequacy and desirability of existing supplies. Now, thanks to Messrs. Sovacool, Sidortsov, and Jones, we have a thoughtful, comprehensive assessment of this important topic."

Michael Klare, Five College Professor of Peace &
World Security Studies, Hampshire College, USA

"A brilliant and much-needed contribution to one of the most pressing issues of our time: meeting global energy needs in ethically defensible ways. Sovacool and his co-authors have distinguished themselves yet again. A must-read for anyone interested in energy."

Kristin Shrader-Frechette, O'Neill Family Endowed
Professor, University of Notre Dame, USA

"This book's perspective is a vital one in the age of climate change; it will become more vital as the impacts of our energy choices harm more people in the future, especially the world's poor. It will be essential reading for anyone interested in the role of energy in the modern world, especially those concerned about the resulting injustices and how to reduce them."

Paul G. Harris, Chair Professor of Global and
Environmental Studies, Hong Kong Institute of Education

"*Energy Security, Equality and Justice* proposes the novel and important idea that energy security should be construed widely to encompass principles derived from modern theories of distributive justice, procedural justice, and cosmopolitan justice. This is a provocative and pathbreaking book that permits us to think about the debate over the transition to a low-carbon energy system in terms of current energy injustices whether human rights abuses and energy-related civil conflicts or the social and class character of energy poverty. A timely and important book."

Michael Watts, Class of 63 Professor at the
University of California, Berkeley, USA

Energy Security, Equality, and Justice

Benjamin K. Sovacool,
Roman V. Sidortsov, and
Benjamin R. Jones

LONDON AND NEW YORK

First published 2014
by Routledge
2 Park Square, Milton Park, Abingdon, Oxon OX14 4RN

and by Routledge
711 Third Avenue, New York, NY 10017

Routledge is an imprint of the Taylor & Francis Group, an informa business

© 2014 Benjamin K. Sovacool, Roman V. Sidortsov, and Benjamin R. Jones

The right of Benjamin K. Sovacool, Roman V. Sidortsov, and Benjamin R. Jones
to be identified as author of this work has been asserted by him/her in accordance
with sections 77 and 78 of the Copyright, Designs and Patents Act 1988.

British Library Cataloguing-in-Publication Data
A catalogue record for this book is available from the British Library

Library of Congress Cataloging-in-Publication Data
Sovacool, Benjamin K.
   Energy security, equality and justice / Benjamin K. Sovacool, Roman V.
Sidortsov, Benjamin R. Jones.
      pages cm
   Includes bibliographical references and index.
1. Power resources—Moral and ethical aspects.   2. Energy policy—Moral
and ethical aspects.   3. Distributive justice.   I. Sidortsov, Roman V.
II. Jones, Benjamin R.   III. Title.
TJ163.2.S692   2014
174'.933379—dc23
2013025436

ISBN: 978-0-415-81519-2 (hbk)
ISBN: 978-0-415-81520-8 (pbk)
ISBN: 978-0-203-06634-8 (ebk)

Typeset in Garamond
by Apex CoVantage, LLC

Printed and bound by CPI Group (UK) Ltd, Croydon, CR0 4YY

Contents

Analytical table of contents

Acronyms and abbreviations

$	refers to United States dollars unless otherwise noted
AEC	U.S. Atomic Energy Commission
BTC	Baku-Tbilisi-Ceyhan oil pipeline
BTUs	British Thermal Units
CBDR	common but differentiated responsibilities
CIA	U.S. Central Intelligence Agency
CNPC	Chinese National Petroleum Corporation
CO_2	carbon dioxide
DECC	U.K. Department of Energy and Climate Change
DOE	U.S. Department of Energy
EIA	environmental impact assessment
EITI	Extractive Industries Transparency Initiative
EPA	U.S. Environmental Protection Agency
EJ	exajoules
EU	European Union
FPIC	free, prior, and informed consent
GDP	gross domestic product
GRP	gross regional product
GJ	gigajoules
GHG	greenhouse gas
GW	gigawatt
GWh	gigawatt-hours
IAEA	International Atomic Energy Agency
IAP	indoor air pollution
IEA	International Energy Agency
IMF	International Monetary Fund
ITDB	IAEA's Incident and Trafficking Database
kW	kilowatts
kWh	kilowatt-hour
LNG	Liquefied Natural Gas
MIT	Massachusetts Institute of Technology
MTVF	mountaintop mining with valley fill operations
MW	megawatt

MWh   megawatt-hours
NASA   U.S. National Aeronautics and Space Administration
NERSA   National Energy Regulator of South Africa
NGO   nongovernmental organization
NO_x nitrogen oxides
OECD   Organization of Economic Cooperation and Development
OPEC   Organization of Petroleum Exporting Countries
PLN   Perusahaan Listrik Negara
PM   particulate matter
PSI   Paul Scherrer Institute
PUCHA   Public Utility Holding Company Act
SO_2 sulfur dioxide
T&D   electric transmission and distribution
UK   United Kingdom
UN   United Nations
US   United States
USGS   U.S. Geological Survey
WHO   World Health Organization

Illustrations

Figures

Tables

One that merely knows right principles is not equal to one that loves them.
— Confucius

Acknowledgments

The Energy Security and Justice Program at Vermont Law School's Institute for Energy and the Environment investigates how to provide ethical access to energy services and minimize the injustice of current patterns of energy production and use. It explores how to equitably provide available, affordable, reliable, efficient, environmentally benign, proactively governed, and socially acceptable energy services to households and consumers. One track of the program focuses on lack of access to electricity and reliance on traditional biomass fuels for cooking in the developing world. Another track analyzes the moral implications of existing energy policies and proposals, with an emphasis on the production and distribution of negative energy externalities and the impacts of energy use on the environment and social welfare.

This book is one of three produced by the Program. The first, *Energy Security, Equality, and Justice,* maps a series of prominent global inequalities and injustices associated with modern energy use, and presents the affirmative and prohibitive justice principles. The second, *Energy and Ethics: Justice and the Global Energy Challenge,* presents a preliminary energy justice conceptual framework and examines eight case studies illustrating countries and communities that have overcome energy injustices. The third, *Global Energy Justice: Principles, Problems, and Practices*, matches eight philosophical justice ideas with eight energy problems, and examines how these ideals can be applied in contemporary decision making.

The authors are most grateful to Professor Aleh Cherp and the Central European University in Budapest, Hungary, for an Erasmus Mundus Visiting Fellowship with the Erasmus Mundus Master's Program in Environmental Sciences, Policy, and Management (MESPOM), which has supported elements of the work reported here. We must state that any opinions, findings, and conclusions or recommendations expressed in this book are our own, however, and do not necessarily reflect the views of Vermont Law School, the Institute for Energy and Environment, Aarhus University, Cambridge University, or Central European University.

More specifically, and lastly, messengers Sovacool, Sidortsov, and Jones thank Stefanie Sidortsova for truly excellent copyediting. Professor Sovacool thanks his family – Lilei, Ethan, Zachary, and Cooper – for their continued love and incredible patience. Mr. Sidortsov expresses his gratitude to Stefanie and Max for tolerating his absence on the weekends, to Svetlana Sidortsova and Valeriy Sidortsov for the wonderful childhood memories, and to Roxana and Dean Bacon for keeping his head straight. Mr. Jones wishes to thank Aimee, Adelaide, and Marguerite for their patience and love.

Dr. Benjamin K. Sovacool is Director of the Danish Center for Energy Technology at AU-Herning and a Professor of Business and Social Sciences at Aarhus University in Denmark. He is also Associate Professor of Law at Vermont Law School and Director of the Energy Security and Justice Program at their Institute for Energy and the Environment. Professor Sovacool works as a researcher, consultant, and teacher on issues pertaining to renewable energy and energy efficiency, the politics of large-scale energy infrastructure, designing public policy to improve energy security and access to electricity, and building adaptive capacity to the consequences of climate change. He is a Contributing Author to the Intergovernmental Panel on Climate Change's (IPCC) forthcoming Fifth Assessment (AR5), and a former Eugene P. Wigner Fellow at the Oak Ridge National Laboratory, as well as the recipient of large research grants from the MacArthur Foundation and Rockefeller Foundation, among others. He has repeatedly consulted for the Asian Development Bank, United Nations Development Program, and United Nations Economic and Social Commission for Asia and the Pacific on energy poverty, governance, and security issues. He is also the author of more than 250 peer-reviewed academic articles, book chapters, and reports and the author, coauthor, editor, or coeditor of 16 books, including *The Routledge Handbook of Energy Security* and *The National Politics of Nuclear Power,* both published with Routledge. He received his PhD in science and technology studies from the Virginia Polytechnic Institute & State University in Blacksburg, Virginia.

Roman V. Sidortsov is a Senior Global Energy Fellow at the Institute for Energy and the Environment at Vermont Law School in the United States, where he teaches oil and gas development and renewable energy courses. He is also a doctoral researcher at the Scott Polar Research Institute at the University of Cambridge in the United Kingdom. Mr. Sidortsov has taught at Irkutsk State Academy of Law and Economics in Russia and at Marlboro College Graduate School's Managing for Sustainability program in the United States. Prior to returning to academia, Mr. Sidortsov practiced law in Russia as in-house counsel for an American nonprofit organization, and in the United States as a transactional attorney. His research focuses on legal and policy issues related to the development of environmentally sustainable energy systems, risk governance in the oil and gas sector, and Arctic offshore oil and gas exploration and extraction, with a special emphasis on the Russian Federation, Norway, and the United States. Mr. Sidortsov serves as a member of the U.S. Academic Team in the Energy Law Partnership of the U.S.-Russia

Bilateral Presidential Commission's Energy Working Group and a member of the Extractive Industries Working Group at the Arctic Centre, University of Lapland. He received his first law degree (Bachelor's and Master's) in the Russian Federation from Irkutsk State University and his Juris Doctor and LL.M degrees from Vermont Law School.

Benjamin R. Jones is a Senior Global Energy Fellow at the Institute for Energy and the Environment at Vermont Law School in the United States, where he teaches graduate courses on energy law and policy and climate change adaptation. He is also a PhD candidate at the University of Victoria in the Faculty of Law. His research looks at the implications of property rights and international investment agreements for the development of domestic energy policy, with a focus on Canada and the European Union. Mr. Jones's original academic training was in philosophy, which he studied at Concordia University and McGill University in Montreal, Canada, and at the Katholieke Universiteit Leuven in Belgium. After finishing his studies, he obtained his commercial pilot's license and moved to northern Canada, where for almost a decade he flew bush planes and worked as an administrator on an Indian reserve. Mr. Jones received his Juris Doctor from Vermont Law School and an MPhil in Environmental Policy from the University of Cambridge.

1 The global energy system beyond technology and economics

Beyond a certain point, more energy means less equity.
                    – Ivan Illich, *Energy and Equity* (1974, p. 45)

Unveiling our not-so-shiny energy world

It is hard to downplay the importance of the energy system to modern society. Patients plan to have surgeries without any regard for availability of electric power. Most families can choose among several transportation options to reach their vacation destination. A homemaker can start cooking dinner by simply turning on the stove. In fact, most people do not put much thought into where their energy comes from and what kind of social, political, economic, and environmental ripples it causes before entering their gas tanks and power outlets. The majority of Americans, Germans, Norwegians, Japanese, and citizens of other Organization of Economic Cooperation and Development (OECD) countries take access to reliable and affordable energy for granted: a doctor expects the lights to turn on when she flips the switch in the operating room; a family travelling to see its out-of-state counterparts is certain that they can refuel at the nearest advertised gas station; a homemaker knows how long it will take him to make a roast dinner at 325 degrees Fahrenheit. Moreover, many of us have underscored our indifference toward the origin and the impact of the energy we use through our enjoyment of products and traditions such as "muscle cars" and neighborhoods clad in Christmas lights.

However, global energy trends do not inspire long-term confidence, and, as Ivan Illich suggests with the introductory quote, sometimes consuming more energy involves even greater levels of inequity and inequality. Not only might the muscle car and overly-illuminated holiday neighborhood be in danger; the very parts of the world that take affordable and reliable energy supply as a given may no longer be in the same worry-free position. We appear to be drifting directly into a future threatened with climate change, rising sea levels, severe pollution, energy scarcity and insecurity, nuclear proliferation, and a host of other dangers. As India and China, among others, seek to consume more, will the West be willing to make space for them in the context of finite energy resources? Conversely, will developing countries do their part to help limit the stock of greenhouse gas emissions going into the atmosphere? Will everyone work together to increase the distribution of electricity in Africa – the continent likely to be hardest hit by the effects of climate change while having contributed least to the problem?

If the global energy system were to remain configured as it is today until the end of the century, the developed part of the world would have a few decades left to continue enjoying the benefits of hot showers and TV nights without much regard for where these pleasant things come from. However, by the year 2100 the economic losses associated with blackouts and power outages would exceed more than $1 trillion each year.[1] The number of traffic-related deaths would surpass 3.6 million per year, the planet would be home to roughly 6 billion automobiles, and the global economy would consume 252 million barrels of oil per day.[2] Natural and humanitarian disasters associated with climate change, driven largely by greenhouse emissions from the energy sector, would impact billions of people, making entire countries such as the Maldives, Kiribati, Tuvalu, and others uninhabitable.[3] More than 3 billion people would lack access to electricity networks, and more than 6 billion would depend on solid fuels whose use directly threatens their health and prosperity, with more dying from indoor air pollution than malaria, tuberculosis, and HIV/AIDS.[4] Egregious human rights abuses associated with energy production – including the denial of free speech, torture, slavery, forced labor, executions, and rape – would continue to afflict thousands of people around the world.[5]

A few recent studies investigating the global sustainability and security of the energy system have confirmed many of these problems. For instance, one scorecard of energy sustainability performance in the United States found that, despite all of its legislation on energy issues over the past five decades, the country has clearly backslid in almost every indicator from the 1970s to today.[6] A similar survey of 22 countries in the OECD found equivalent results: Denmark, with the best score, led the pack with a "3" when their highest possible score was a "10." A majority of countries did poorly, with 13 countries scoring below zero, implying that their energy security had worsened from 1970 to 2007.[7] A final assessment of countries in the Asia-Pacific utilized 20 distinct indicators spread broadly across the areas of energy supply, energy affordability, efficiency and innovation, environmental stewardship, and governance.[8] Japan and Brunei scored favorably on slightly more than only 50 percent of the metrics and the third-best performer, the United States, scored favorably only for one-third of its metrics.

Put succinctly, the current energy system resembles a two-faced Janus. One face admires the seemingly worry-free energy past with its cheap electricity, transportation and heating fuels. The other cringes at the future as the system spews hazardous material into our air, land, and water, leading to significant deterioration of the human and natural environment. The current energy system is the single biggest contributor to climate change. It costs customers billions of dollars every year in blackouts and interruptions in fuel supply, and leaves billions of people without even basic access to modern energy services. Current patterns of energy production and use threaten the very vitality and existence of entire countries, and lead to extreme social and environmental costs for marginalized communities and countries with the least adaptive capacity and resilience to address them.

Clearly, such a troubling outlook makes global energy security and access one of the central justice issues of our time, with profound implications for our notions of social welfare, virtue and equity. Any self-respecting society must grapple with the energy-related injustices that weaken its social fabric, degrade the environment that sustains it, and burden the economy that supports it. As Martin Luther King once put it, "Injustice anywhere is a threat to justice everywhere."

The problem, however, is that by and large people want low-cost access to energy, they want reliable energy, and they want a cleaner environment. But these problems are inextricably linked in a continuous and seemingly perpetual catch-22. Conventional coal may be cheap, but its combustion also generates both greenhouse gases and smog emissions. Many environmental problems could be solved if energy use was cut in half, but doing so would also devastate the global economy. And reliability would benefit, say, from larger operating reserves for electricity generation, but these reserves would increase both the cost of delivered electricity and the extent of environmental waste. Yet despite these tradeoffs, the vast majority of energy analysts and policymakers approach their craft from one (economic) or, at best, two (economic and reliability) angles. Other dimensions are often left out of the policy-making process and are treated with neutrality or (at worst) moral apathy.[9]

The concept of "energy justice" creates a more comprehensive and, potentially, better way to assess and resolve these dilemmas. Energy justice, as we will elaborate in Chapter 2, consists of two key principles:

- A *prohibitive principle,* which states that "energy systems must be designed and constructed in such a way that they do not unduly interfere with the ability of people to acquire those basic goods to which they are justly entitled," and
- An *affirmative principle,* which states that "if any of the basic goods to which people are justly entitled can only be secured by means of energy services, then in that case there is also a derivative entitlement to the energy services."

These two principles are premised on the notion that energy serves as a material prerequisite for many of the basic goods to which people are entitled. They also recognize that the externalities associated with energy systems often interfere with the enjoyment of such fundamental goods as security and welfare. Our principles apply to decisions about energy systems and resources, especially when choosing between two competing courses of action. They acknowledge that the structuring of energy systems has profound ramifications for human societies, providing historically unprecedented benefits for some, and taking from others the possibility of living a life of basic human dignity. The Harvard philosopher Michael J. Sandel wrote that "to ask whether a society is just is to ask how it distributes the things we prize – income and wealth, duties and rights, powers and opportunities, offices and honors. A just society distributes these goods in the right way; it gives each person his or her due."[10] Energy justice, therefore, recognizes that energy needs to be included within the list of things we prize; how we distribute the benefits and burdens of energy systems is preeminently a concern for any society that aspires to be just. Thus, we believe that an understanding of philosophy, law, and ethics, along with politics, economics, and history, is elemental in ensuring that decision-makers comprehend the depth and range of their actions concerning energy production and use.

Statistical snapshot of the global energy system

A conversation about the justice implications of energy use must begin by describing the system in place to convert and distribute energy. The global energy system is massive in its output, expansive in its reach, one-dimensional in its primary fuel source, and expensive to build and maintain. As Figure 1.1 shows, humanity used about 550 exajoules (EJ) of energy in 2010. Of this total, 80 percent was provided by fossil fuels, a mere 11 percent by biomass and biofuels, 5.5 percent from nuclear energy, 2.2 percent from hydroelectric

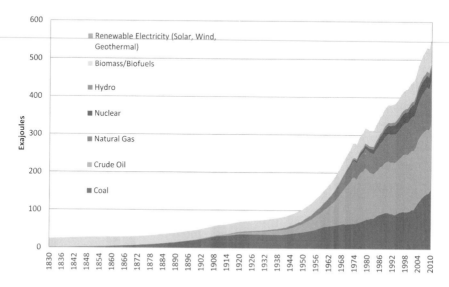

Figure 1.1 Worldwide energy production, 1830 to 2010 (exajoules)

Source: International Energy Agency.

dams, and less than 0.4 percent from other renewable electricity sources. Although 550 EJ is a staggering amount of energy, demand will grow 45 percent globally between 2010 and 2030, and this amount will almost triple by the end of the century. Between 2011 and 2035, the International Energy Agency expects $38 trillion to be invested in the global energy sector.[11]

When put into perspective, the dimensions of the current global energy system, as well as its addiction to fossil fuels, are remarkable. For example, when the Vulcan Street Plant, the first Edison hydroelectric power station, began producing electricity on September 30, 1882, in Appleton, Wisconsin, its capacity was about 12.5 kilowatts (kW).[12] The capacity of the Three Gorges Dam Hydropower Plant, currently (as of 2013) the largest in the world, exceeds that of the Vulcan Street Plant 1,456,000 times at 18,200 MW.[13] The first airliner, Igor Sikorsky's *Ilya Muromets*, had four 100 horsepower engines. When the airplane made its debut flight in 1913, it was considered a true engineering marvel equipped with a passenger saloon and an onboard bathroom.[14] An Airbus A380 is also powered by four engines, each certified at 70,000 pounds of thrust, roughly equivalent to 31,818 horsepower.[15] In typical three-class configuration, the A380's two decks can seat 525 passengers[16] (853 in a single class configuration), accommodated by individual in-flight entertainment systems and 17 bathrooms during a typical 15-hour transcontinental flight.[17] At the beginning of the twentieth century, a successful American farmer generated about 5 kW of power from a team of six horses while plowing his field. A century later, an average American farmer sits atop a diesel-powered tractor that boasts 250 kW. And he does so in an air-conditioned cabin while enjoying listening to his favorite country-western tunes via satellite radio. From 1900 to 2000 the population of the earth quadrupled from 1.6 billion to 6.1 billion, but annual average supply of energy per capita grew *even more,* from 14 Gigajoules (GJ) in 1900 to roughly 60 GJ in 2000. Over this period, energy consumption in the US more than tripled, quadrupled in Japan, and increased by a multitude of 13 in China.

These trends of energy consumption highlight the accelerating nature of human life. Modern economies are driven by the growth imperative. To satisfy it, people are encouraged to make more, export more, and buy more. Our ever-expanding economy, propelled by "needs" that we did not know we had even a few years ago, can no longer run on the energy derived directly from the sun. Simply put, human and animal muscle power, flowing water, wind, and wood alone are insufficient to run the modern world. It would take 13.3 billion people – almost twice as many as exist on Earth – to naturally power America's electricity grid for an hour.[18]

Instead of exploring better ways in which to harness the power of the sun and questioning the blind need for more energy, we have taken a shortcut. It came in the form of fossil fuels, the product of millions of years of the sun's work compacted and refined by millions of years of anaerobic decomposition. This explains why the global per capita use of hydrocarbons, for example, surged 800-fold from 1750 to 2000 and 12-fold, again, from 1900 to 2000.[19] Accelerated economies need highly concentrated, high-octane fuel to stay in the growth race. People no longer have the time to look up at and appreciate our ultimate power source. Instead, they keep their heads down looking for the remains of long-dead plants and animals to put in gas tanks or stock fire chambers.

Accordingly, the modern global energy system encompasses many fuel chains extending in a number of directions. It can be viewed not only as a compilation of national energy systems but also as comprised of smaller networks united by certain fuels. A marquee example of such a network is a coal system that incorporates coal mines, railways, coal-fired power plants, and an electrical transmission and distribution network. The extractive sector provides raw fuels such as unwashed coal, unprocessed natural gas, and crude oil from hundreds of thousands of mines and production wells. The processing sector converts these raw materials into usable fuels at coal handling and preparation plants and refineries. Finally, the power sector converts them into electrons. Yet the overlap extends even further as iron, copper, rare earth elements, bauxite, and other metals are needed to manufacture power plants, transmission lines, and other infrastructure.[20] Although the global energy system can be dissected and examined from many different angles, we will briefly look at three interconnected yet distinct sectors for a more detailed picture: electricity, transport, and heating and cooking (thermal needs).

Electricity

Electricity occupies a special place in the global energy system. Unlike the other two sectors reviewed in this section, electricity is not purely characterized by its end use. It is truly a universal good and service that extends into the transportation and thermal sectors. Electricity now powers not only lights, televisions, refrigerators, and cooktops, but also vehicles, movable sidewalks, neon signs, and electric fireplaces. Dependence on electricity has grown significantly due to its easily adjustable flow, easy and instantaneous access, and minimal disturbance to the surrounding environment at the point of use. Humans have also become more reliant on information and technology and electronic means of communication such as the Internet and mobile phones, which have led to a dramatic uptake in the manufacturing of information storage, telecommunications, and electronics. As a result, our high-tech world has become more dependent on the energy needed to power such devices.[21]

The global electrical power system is a vast network consisting of approximately 170,000 generators producing electricity at over 75,000 power plants.[22] In 2010, the system produced 20,225 billion megawatt-hours (MWh) of electricity.[23] Almost 67 percent of the world's electricity or 13,473 billion MWh came from fossil fuels; 20 percent or 4,154 billion MWh came from renewable sources, with hydro providing almost 17 percent or 3,402 billion MWh; and nuclear power plants generated 13 percent or 2,620 billion MWh.[24] In comparison, in 1950, the share of fossil fuels in electricity production was 10 percent, and in 1900, less than 2 percent.[25]

As of 2012, the global electrical system transferred power through approximately 4 million miles of transmission and distribution lines. The world's longest transmission line, located in the Democratic Republic of the Congo, extends more than 1,056 miles and cost $900 million to build in 1982. In the United States, almost 20,000 power plants and half a million miles of high-voltage transmission lines serve the world's largest economy.[26] Thirteen hundred coal mines, 125 nuclear waste storage facilities, and 410 underground natural gas storage fields, along with hundreds of millions of distribution points, transformers, and electric motors complete the country's gargantuan electric power sector.[27]

The electric power sector is also the most capital-intensive sector of many economies. The total investment in the sector for the United States represents about 10 percent of all sunk investment nationwide. In 2007, expenditures on electricity reached 3.2 percent of the nation's annual GDP, or $355 billion.[28] The electricity sector in the United States is dotted with firms of all shapes and sizes ranging from municipal utilities to consulting firms. The "multiplication effect" is especially prevalent in the nuclear industry, where enrichment facilities, uranium mines, fuel cladding facilities, uranium mills, permanent waste repositories, temporary waste sites, and research laboratories cater to each phase and subphase of the fuel cycle.

From the electric kettle to the smart phone – electricity has transformed nearly every aspect of daily life in the modern world. It has and will continue to improve the productivity and quality of life of people around the world. Because of its extraordinary versatility, electricity can be put to almost every use. From heating to lighting and from moving freight to enabling people to communicate over long distances, electricity powers the world.

Transport

The introduction of mobile engines and inexpensive liquid fuels has enhanced personal mobility and facilitated new modes of transport. The number of mass-produced motorized vehicles jumped from a few thousand in 1900 to more than 700 million in 2000 (and about 1 billion in 2012), along with a notable increase in vacation travel and nonessential trips. In 1900, for instance, there were only 8,000 cars in the United States and just 15 kilometers of paved roads.[29] The mobility of people has been matched by the increasing movement of goods and services as trade and commerce have accelerated. In 2000, trade alone accounted for 15 percent of global economic activity, most of it through diesel-powered rail, freight, trucks, and tankers.

The rise of the automobile is often characterized as one of the great achievements of the twentieth century. During the first half of the century, the gasoline-powered vehicle evolved from a fragile, cantankerous, and faulty contraption to a streamlined, reliable, fast, luxurious, and widely affordable product.[30] These automotive engineering feats were enhanced by the creation of interstate highway systems and urban infrastructure that have offered many people unprecedented mobility.

The transport system that resulted must deliver about 19 million barrels of oil per day to the United States alone, which accounts for 21 percent of the global demand of 87.6 million barrels of oil per day (as Table 1.1 shows). This oil is processed by roughly 1,000 refineries and the bulk of it eventually transported to almost 1 million gasoline stations. Its final destination: the fuel tanks of the world's roughly 1 *billion* automobiles, which drive on 11.1 million miles of paved roads – enough to drive to the moon and back 46 times.[31] These roads require $200 million of maintenance per day, and constitute a paved area equal to all arable land in Ohio, Indiana, and Pennsylvania. The transport system also creates seven billion pounds of un-recycled scrap and waste each year. The United States has a voracious appetite for oil, with consumption having climbed 25 percent since the mid-1980s. China is also learning to drink deep, and may be using an equivalent amount to the United States in a relatively short timeframe.

Consumers worldwide have precious little choice surrounding their fuels for transport: the United States relies on crude oil to meet 94 percent of its transportation demand; for the world, the number rises to 96 percent (electricity, biofuels, and natural gas meet the rest).[32] In the developed world, the car maintains its enduring stranglehold over public transportation options. Statistics agencies in many North American states and provinces report that over 80 percent of their workforce commutes by car. In other words, gasoline vehicles have become a transportation monoculture: cars and petroleum dominate. Automobile domination is evidenced by the fact that public transport accounts for only about 3 percent of passenger travel in the U.S.,[33] and rail transport accounts for less than 1 percent.[34] Gasoline domination is evidenced by the fact that today non-petroleum fuels (including electricity, biofuels, and natural gas) account for only 4 percent of U.S. transportation fuel consumption (up from 2 percent in 2007).[35] The motorization of America has resulted in more than one auto for every licensed U.S. driver.

In the developing world, dirt-cheap cars are gaining momentum in a marketplace of billions. Throughout India, China, and Indonesia, middle-class families are laden 3 or 4 to a single scooter, choking through diesel-exhaust on overcrowded roads. Although 85 percent of the world's population does not have access to a car, they aspire to car ownership, especially residents of the rapidly growing South and East Asia nations. In China, the conventional vehicle fleet is expected to grow tenfold from 37 million in 2005 to 370 million by 2030.[36] In total, within a couple of decades, the world is projected to have 2 billion gasoline-powered automobiles – twice the number that currently exists.[37] If China continues its car-centric development model, it could by itself add another billion cars by the end of the century.

Heating and cooking

Though electricity accounts for roughly 17 percent of global final energy demand, low-temperature heat accounts for about 44 percent. Globally,

Table 1.1 Selective global petroleum statistics, 2011

Country name	Petroleum consumption (thousand barrels per day)	Petroleum production (thousand barrels per day)	CO$_2$ emissions from petroleum consumption (million metric tons)	Population	Petroleum consumption per capita (gallons per day)	Petroleum production per capita (gallons per day)	CO$_2$ emissions from petroleum consumption (pounds per day)
World	87,065.30	87,328.90	11,407.68	6,974,242,787	0.53	0.53	9.88
North America	23,382.50	16,692.90	2,855.72	–	–	–	–
Canada	2,289.30	3,597.30	284.45	34,483,975	2.79	4.38	49.82
United States	18,949.40	10,135.60	2,299.00	311,591,917	2.55	1.37	44.56
Central & South America	6,914.70	7,857.30	915.80	–	–	–	–
Brazil	2,793.00	2,685.20	372.91	196,655,014	0.60	0.57	11.45
Europe	15,082.70	4,269.00	2,002.88	–	–	–	–
France	1,791.50	75.90	237.16	65,433,714	1.15	0.05	21.89
Germany	2,400.10	161.90	306.06	81,797,673	1.23	0.08	22.60
United Kingdom	1,607.90	1,166.80	221.59	62,744,081	1.08	0.78	21.33
Eurasia	4,271.10	13,332.30	509.27	–	–	–	–
Russian Federation	2,725.00	10,239.20	337.59	142,960,000	0.80	3.01	14.26
Middle East	7,905.00	26,873.90	1,063.67	–	–	–	–
Africa	3,505.00	9,367.40	481.01	–	–	–	–
Asia & Oceania	26,544.30	8,936.30	3,579.33	–	–	–	–
Australia	1,023.10	526.40	139.87	22,323,900	1.92	0.99	37.84
China	8,924.00	4,347.00	1,279.79	1,344,130,000	0.28	0.14	5.75
India	3,426.00	995.80	449.82	1,241,491,960	0.12	0.03	2.19
Japan	4,480.50	136.30	523.15	127,817,277	1.47	0.04	24.72

Source: U.S. Energy Information Administration and the World Bank.

this means that people use more energy for heating, cooking, and thermal comfort – typically by burning woody biomass – than for any other purpose.[38]

For instance, the majority of households in the developing world – about three out of every four – rely on traditional stoves for their cooking and heating needs. Traditional stoves range from three-stone open fires to substantial brick and mortar models and ones with chimneys.[39] Because these stoves are highly inefficient (as much as 90 percent of their energy content is wasted), they require a significant amount of fuel – almost two tons of biomass per family per year. These consumption patterns strain local timber resources and can cause "wood fuel crises" when wood is harvested faster than it is grown.[40] The burning of this biomass also exacerbates climate change. Households in developing countries burn about 730 tons of biomass annually, which translates into more than 1 billion tons of carbon dioxide.[41] By comparison, Japan, currently the fifth-largest carbon dioxide emitter in the world, emitted 1.2 billion tons of carbon dioxide in 2008.[42]

Billions of people consequently rely on wood, charcoal, and other biomass fuels to meet their daily energy needs. Such fuels comprise 40 to 60 percent of total energy consumption for many communities, and household cooking is responsible for 60 percent of total energy use in sub-Saharan Africa (and exceeds 80 percent in some countries). Poor families spend one-fifth or more of their income on wood and charcoal, devote one-quarter of household labor collecting fuel wood, and then suffer the life-endangering pollution that results from inefficient combustion.[43]

Energy et al. defined

Before embarking on the rest of our journey, it is worthwhile to define a few concepts repeatedly encountered throughout the book's chapters. We start with the largest building block that defines the very foundation of this book: energy.

For scientists and engineers, the term "primary energy" means the energy "embodied" in natural resources, such as coal, crude oil, natural gas, uranium, and even sunlight, wind, geothermal heat, or falling water, which may be mined, stored, harnessed, or collected but not yet converted into other forms of energy. Sometimes analysts use the term "end-use energy," to refer to the energy supplied to the consumer at the point of end use, such as kerosene, gasoline, or electricity, delivered to homes and factories. The phrases "useful energy," "useful energy demands," and "energy services" are what we are most interested with in this study, and refer to what "end-use energy" is transformed into: heat for a stove or mechanical energy for air circulation. "Energy services" are often measured in units of heat, work, or temperature, but these are in essence surrogates for measures of satisfaction experienced when human beings consume or experience them. Energy services can thus be regarded as the benefits that energy carriers produce for human well-being.[44]

In their work after the energy crises of the 1970s, behavioral psychologists Stern and Aronson identified at least three ways of looking at energy from the perspective of different disciplines. The scientific view of physicists and engineers frames energy as a property of heat, motion, and electrical potential, measurable in joules and British Thermal Units (BTUs). According to this view, energy is neither produced nor consumed, quantity is always conserved, quality is always declining, and correct policy is a matter of understanding thermodynamics and physics.[45] From a thermodynamic perspective, human beings are merely complex organisms for processing energy: There is no more fundamental issue for humans than the energy that fuels their biological existence.[46]

Stern and Aronson also described an "economic view" that sees energy as a commodity – or a collection of commodities such as electricity, coal, oil, and natural gas – traded on the market. This view emphasizes the value of choice for consumers and producers and assumes that the marketplace allocates choices efficiently. According to this view, when prices rise, fuel substitutes will be found, and inequities arise only through irrational behavior. Correct policy is a matter of analyzing transactions between buyers and sellers and minimizing the external costs of these transactions. The economic view of energy has become perhaps the most dominant one recently, with strong undertones of competition and "efficient" markets promoted in the energy sector even if regulators fail to recognize market failures.[47]

Finally, Stern and Aronson described an "ecological view" which rejects framing energy in scientific or economic terms, and instead classifies energy resources as renewable or nonrenewable, clean or polluting, and inexhaustible or depletable, to emphasize their environmental context. This view prioritizes the values of sustainability, frugality, and future choice. Correct policy is a matter of recognizing that energy resources are finite and interdependent and that present use engenders significant costs to future generations.

We draw upon each of the aforementioned views – scientific, economic, and ecological – and offer the following definition of energy systems: the sociotechnical systems in place to convert energy fuels and carriers into services – thus not just technology or hardware such as power plants and pipelines, but also other elements of the "fuel cycle" such as coal mines and oil wells in addition to the institutions and agencies, such as electric utilities or transnational corporations, that manage the system, as well as the households and enterprises that consume or put that energy to work.

The term "energy security" plays a special role in this book. Yet we do not use this term as "the uninterrupted availability of energy sources at an affordable price," which is the way the International Energy Agency (IEA) defines it. We also do not identify it as "security of demand," which is the standpoint shared by nations whose economic well-being depends on energy exports.[48] We instead define energy security as "equitably providing available, affordable, reliable, efficient, environmentally benign, proactively governed, and socially acceptable energy services to end users." This conception of energy security comes from a literature review of the academic literature on energy security[49] as well as research interviews with energy experts[50] and surveys of energy end users.[51] This emerging multidimensional view of energy security acknowledges that transforming energy systems is both a scientific and social feat.

Furthermore, we use the terms "conventional energy sources" and "conventional energy systems" frequently throughout the book. We define the former as "sources accounting for the vast majority of global energy production." We include all fossil fuels, conventional and not (e.g., shale gas and tight oil), nuclear fuels, and large hydroelectric sources in this category. The category of non-conventional energy resources is dominated by non-hydro renewables such as geothermal, wind, biomass, and solar, as well as renewable liquid fuels.[52] The term "conventional energy system" refers to the infrastructure used to produce, transport, process, and convert the majority of conventional energy sources. These systems are characterized by their centralized approach (e.g., a 1 gigawatt [GW] nuclear power plant serving millions of customers), large size (for example, an oil pipeline spanning the width of the United States and capable of transporting 500,000 barrels of oil a day), and high capital cost (for example, Royal Dutch Shell spent over $4 billion on exploration off the Alaska coast but managed to drill only two "pilot" holes).[53] The non-conventional energy systems include distributed generation, microgrids, combined heat and power, energy conservation, and energy efficiency.

A great deal of discussion in the ensuing chapters involves juxtaposing developed and developing countries. Several organizations have produced classifications placing the world's nations in different camps. We employed the International Monetary Fund's (IMF) classification from its World Economic Outlook.[54] According to the IMF, the developed world consists predominately of OECD countries and a few wealthy Asian nations.[55] A few notable newcomers who "graduated" to the rank of advanced economies include Israel, Singapore, South Korea, and the Czech Republic.[56]

We end this section with the concept that serves as the second-largest building block in the foundation of this book: justice. Admittedly, justice is a difficult notion to tie down. As philosopher Scott Gordon puts it, "justice is the central concern of law and jurisprudence and a large part of the social sciences, and it is also a major one of philosophy, theology, and the arts."[57] For the ancient Greeks, justice involved living a virtuous life; for modern libertarians, it is about minimizing government intervention and control over individual choices; for social philosophers, it can be about equality and welfare. Some believe justice is inherently tied to the law, and to retributive or preventative orders made by a judge or jury or an impartial, official authority like Congress. Others believe justice concerns individual liberty, and the ability of a citizen to pursue freely – and hopefully realize – his or her individual desires. Many modern notions of justice focus on the concept of "fairness" and attempt to create the conditions for fair social structures, which in turn produce a fair distribution of goods and services.

But what is a more concrete definition of justice? John Rawls opened his celebrated *Theory of Justice* with the following words:

> Justice is the first virtue of social institutions, as truth is of systems of thought. A theory however elegant and economical must be rejected or revised if it is untrue; likewise laws and institutions no matter how efficient and well-arranged must be reformed or abolished if they are unjust. Each person possesses an inviolability founded on justice that even the welfare of society as a whole cannot override. For this reason justice denies that the loss of freedom for some is made right by a greater good shared by others. It does not allow that the sacrifices imposed on a few are outweighed by the larger sum of advantages enjoyed by many. Therefore in a just society the liberties of equal citizenship are taken as settled; the rights secured by justice are not subject to political bargaining or to the calculus of social interests.[58]

As Chapter 2 will explain in greater detail, we develop the concept of energy justice, combining for the purpose the disciplines of political philosophy and ethics with energy policy. We start with four assumptions about social justice and energy systems, and from them arrive at two principles – our *prohibitive principle* and our *affirmative principle* – which encapsulate our conception of energy justice. In sum, energy justice contemplates the possibility of a global energy system that fairly disseminates both the benefits and costs of energy services, and one that has representative and impartial energy decision-making. It involves the following key elements:

- Costs, or how the hazards and externalities of the energy system are imposed on communities unequally, often the poor and marginalized;
- Benefits, or how access to modern energy systems and services are highly uneven;
- Procedures, or how many energy projects proceed with exclusionary forms of decision-making that lack due process and representation.

A world that incorporated the principles of energy justice would be one that promoted social welfare, freedom, equality, and due process for both producers and consumers. It would take care to distribute the environmental and social hazards associated with the energy sector in a way that did not disproportionately burden the poor and vulnerable. It would ensure that access to energy systems and services was equitable. It would guarantee that energy procedures were fair and that stakeholders had access to information and participation in energy decision-making.

Viewing energy justice in this manner is novel in at least two ways. First and foremost, our book's focus on the interplay of justice and practical decision-making is unusual: On the one hand, the issue of justice is often neglected in energy and climate policymaking; on the other, pragmatic decisions about the fundamental technical infrastructure of society are frequently given little or no consideration in the literature of justice theory. Yet, as one Brookings Institution study recently noted:

> Decisions or indecisions today can impose heavy costs on our descendants or, at a minimum, limit the choices they will have. That is why there is an unprecedented need to merge the reality of an international community with the established principle of intergenerational responsibility.[59]

Our book does this, but grounds the discussion of justice in real-world case studies, connecting lofty ideals with empirical examples. People are starting to recognize that the world of energy involves fundamental ethical questions. Thirty years ago, electrons and justice would have seemed like a confusion of disciplines. But now it makes sense.

Second, the book has a comparative, global focus with examples from every continent. This is because energy issues are now global in scope. At one time it may have been appropriate for people to think about energy issues in a wholly regional context – but no longer. Just as the world as a whole has become increasingly interdependent, so the same is true for the world of energy.

A road map of chapters to come

Discourse about energy justice, when it does rarely occur, often echoes the narrative of the standoff between developed and developing nations that has been stalling climate negotiations since the adoption of the United Nations Framework Convention on Climate Change. In sum, developing nations argue that because developed countries fueled their economies by cheap fossil fuels and thus have benefited from present environmental degradation, they should bear the responsibility for mitigating climate change.[60] This equity-based approach serves as the premise for the principle of common but differentiated responsibilities (CBDR). The Rio Declaration gave CBDR the status of one of the cornerstones of international environmental law, capturing it as Principle 7 of the document:

> States shall cooperate in a spirit of global partnership to conserve, protect and restore the health and integrity of the Earth's ecosystem. In view of the different contributions to global environmental degradation, States have common but differentiated responsibilities. The developed countries acknowledge the responsibility that they bear in the international pursuit to sustainable development in view of the pressures their societies place on the global environment and of the technologies and financial resources they command.[61]

It is hard to debate CBDR's equitable grounds. Cheap oil and plentiful coal did transform Great Britain, Germany, the United States, France, Japan, and other wealthy countries into lands of widespread car ownership, affordable air travel, indoor plumbing, and other modern conveniences. And many developing nations were literally left in the dust, which remains a fixture on the dirt roads of Angola, Bolivia, Bangladesh, and Yemen. Yet the dispute about *responsibility* for bringing about a cleaner, healthier, and more equitable world has overshadowed the imperative to *cooperate* to achieve this lofty goal. Meanwhile, billions of people suffer from energy-related injustices every day and billions of lives are thus implicated by the decisions that energy producers and consumers make without taking into account the considerations of justice, equality, and security.

Therefore, we chose to abandon this deficient approach. Instead, we highlight in the pages to come the need for a just energy system by examining instances of energy injustice in five different dimensions. These dimensions are driven by the context in which decisions about energy are made. Whether it is balancing present and future costs and benefits, assessing economic impacts, examining the source of human rights abuses, scrutinizing regional planning, or optimizing energy portfolios, we group the justice-related implications of these decisions in temporal, economic, sociopolitical, geographic, and technological dimensions. These five dimensions, presented after Chapter 2 (which introduces our energy justice framework), also serve as the key themes for our five substantive chapters.

The temporal dimension: Externalities and climate change

Chapter 3 looks at present and future justice implications of environmental externalities caused by the global energy system. The chapter's central theme addresses whether preserving the present energy system is justified in the face of global climate change, resource depletion, mounting levels of nuclear waste, and air pollution. The temporal dimension brings the concepts of justice, equality, and security into the economist-dominated debate regarding the present and future costs of environmental externalities. We maintain that energy analysis and decision-making should go beyond the question of the correct discount rate and include an inquiry as to the distribution of current benefits and present and future costs.

The economic dimension: Inequality, poverty, and rising prices

Chapter 4 focuses on the economic impacts of the global energy system. By and large, we look past the statistical averages, such as Gross Domestic Product, and examine the impacts of energy production and use (negative and positive) from a distributive justice angle. We tie the issue of equitable distribution of reliable and affordable energy services to exclusion from meaningful participation in economic activities. We take special note of the connection between conventional energy systems and narrowing economic opportunities that lower the quality of life for certain segments of developed societies. The chapter concludes with an analysis of the energy sector's systemic deficiency that stems from reliance on finite and increasingly more expensive, fossil fuels.

The sociopolitical dimension: Corruption, authoritarianism, and energy conflict

Chapter 5 is premised on the view that energy systems both reflect and reinforce the structure of political and economic power within a society. We take the reader on a tour of the energy justice implications arising out of the unhealthy relationship the energy industry has with corrupt, abusive governments, and militant governments and political leaders. We show the impact that energy has on democracy, the political process, and human rights. Our inquiry into the sociopolitical dimension of energy injustice starts with the role of energy in breeding corrupt governments and social marginalization, and ends with energy as a cause of international military conflicts.

The geographic dimension: Uneven development and environmental risks

The vivid imagery of mining ghost towns, abandoned farmland overshadowed by the giant cooling towers of a large power plant, flooded pristine rainforest, children studying under the light of the village square, a refinery located in the middle of a poor neighborhood, and streets of lower Manhattan in the aftermath of Hurricane Sandy provide the preview of the injustices discussed in Chapter 6. In this chapter, we look at how energy systems affect people on every geographic scale. We examine the effect of energy systems on uneven development of remote areas, the boom and bust problem of many mining towns, and community degradation and displacement due to energy projects. The chapter ends with an examination of the geographic impact of the worst environmental problem ever faced: climate change.

The technological dimension: Efficiency, reliability, safety, and vulnerability

Chapter 7 differs slightly from the preceding chapters. It focuses on the justice implications that are inherent to conventional energy systems instead of injustices that arise from the impact of these on society, the economy, and the environment. To discover the justice implications "implanted" in conventional energy technologies, we examine whether simply importing conventional systems from developed countries to developing ones would represent a just and equitable solution that would also improve energy security. We scrutinize the record of conventional energy technologies in terms of efficiency, reliability, safety, and vulnerability. The chapter concludes with a discussion about the structural deficiencies of conventional energy technologies (regardless of the system's location in a developing or developed country) and the justice implications that they create.

Although we review energy injustices in the aforementioned five dimensions, this does not mean they exist in isolation. Quite the contrary: It would be hard to find a one-dimensional instance of energy justice. Siting a refinery in a poor urban neighborhood not only contributes to uneven geographic distribution of environmental externalities, it also further marginalizes the affected fraction of society (especially if such a practice is common or pervasive). However, examining the complexity of energy injustice not only helps us understand it better, it also places that injustice in the context of something to be addressed and remedied.

Notes

1. Oracle Utilities, *The Future of Energy,* October 14, 2011, reports that in the EU power outages cost $150 billion euros per year. See http://www.oracle.com/us/industries/utilities/utilities-future-energy-525446.pdf. Also, in the United States the cost of annual blackouts was estimated at about $206 billion. For more on this, see Chapter 7.
2. We have merely tripled the existing externalities from today's levels (as of 2013) to correspond with a trebling in energy demand; see Chapter 3.
3. See Chapter 3.
4. See Chapter 4.
5. See Chapters 5 and 6.
6. M. A. Brown and B. K. Sovacool, "Developing an 'Energy Sustainability Index' to Evaluate Energy Policy," *Interdisciplinary Science Reviews* 32, no. 4 (2007): 335–349.
7. See B. K. Sovacool and M. A. Brown, "Competing Dimensions of Energy Security: An International Review," *Annual Review of Environment and Resources* 35 (2010): 77–108; B. K. Sovacool and M. A. Brown, "Measuring Energy Security Performance in the OECD," in *The Routledge Handbook of Energy Security,* ed. B. K. Sovacool (London: Routledge, 2010), 381–395.
8. B. K. Sovacool, I. Mukherjee, I. M. Drupady, and A. L. D'Agostino, "Evaluating Energy Security Performance from 1990 to 2010 for Eighteen Countries," *Energy* 36, no. 10 (2011): 5846–5853.
9. James M. Buchanan, "The Relatively Absolute Absolutes," paper presented at the annual SEA meeting, Washington, DC, November 22, 1987.
10. Michael Sandel, *Justice: What's the Right Thing To Do?* (New York: Farrar, Straus and Giroux, 2009), 19.
11. International Energy Agency, *World Energy Outlook 2011* (Paris: OECD, 2011), 2. See also International Energy Agency, *Key World Energy Statistics 2011* (Paris: OECD, 2011).
12. American Society of Mechanical Engineers, *Vulcan Street Power,* http://www.asme.org/about-asme/history/landmarks/topics-a-l/electric-power-production-water/-29-vulcan-street-power-plant-(1882) (accessed May 28, 2013).
13. China Three Gorges Corporation, *Three Gorges Project,* http://www.ctgpc.com/benefits/benefits_a_2.php (accessed May 28, 2013).
14. Igor Sikorsky, *Story of the Winged-S: An Autobiography by Igor Sikorsky* (New York: Dodd, Mead & Company, 1938), 4.

15. Rolls-Royce, *Trent 900,* http://www.rolls-royce.com/civil/products/largeaircraft/trent_900 (accessed May 28, 2013). The comparison is made with Rolls-Royce Trent 900 engines.
16. Airbus, *A380,* http://www.airbus.com/aircraftfamilies/passengeraircraft/a380family/a380–800/specifications (accessed May 28, 2013).
17. Lufthansa, *Seat Maps,* http://www.lufthansa.com/uk/en/Seat-maps (accessed May 28, 2013).
18. Assuming that the average human body produces between 60 and 90 Watts (W) of equivalent energy per hour. B. K. Sovacool, *The Dirty Energy Dilemma* (Westport, CT: Praegar, 2008).
19. Charles Hall, Preadeep Tharakan, John Hallock, Cutler Cleveland, and Michael Jefferson, "Hydrocarbons and the Evolution of Human Culture," *Nature* 426 (2003): 318–322.
20. A. Goldthau and B. K. Sovacool, "The Uniqueness of the Energy Security, Justice, and Governance Problem," *Energy Policy* 41 (2012): 232–240.
21. Ibid.
22. V. Smil, "Energy in the Twentieth Century: Resources, Conversions, Costs, Uses, and Consequences," *Annual Review of Energy and the Environment* 25 (2000): 21–51.
23. U.S. EIA, "International Energy Statistics," http://www.eia.gov/cfapps/ipdbproject/IEDIndex3.cfm?tid=2&pid=2&aid=12 (accessed May 28, 2013).
24. Ibid.
25. Ibid.
26. Ibid.; "World's Largest Economies," *CNN Money,* http://money.cnn.com/news/economy/world_economies_gdp (accessed May 25, 2013).
27. Smil, "Energy in the Twentieth Century."
28. Energy Information Administration (EIA), *An Updated Annual Energy Outlook 2009 Reference Case Reflecting Provisions of the American Recovery and Reinvestment Act and Recent Changes in the Economic Outlook,* SR/OIAF/2009–03 (Washington, DC: DOE, 2009): Tables A1 and A8, http://www.eia.doe.gov/oiaf/servicerpt/stimulus/excel/aeostimtab_18.xls.
29. Norman Myers and Jennifer Kent, *Perverse Subsidies: How Tax Dollars can Undercut the Environment and the Economy* (Washington, DC: Island Press, 2001).
30. W. M. Shields, "The Automobile as an Open to Closed Technological System: Theory and Practice in the Study of Technological Systems" (PhD diss., Virginia Polytechnic Institute and State University, 2007).
31. Goldthau and Sovacool, "The Uniqueness of the Energy Security."
32. International Energy Agency, *Key World Energy Statistics 2011.*
33. Federal Highway Administration, http://www.fhwa.dot.gov/ohim/tvtw/tvtpage.cfm.
34. The National Railroad Passenger Corporation, known as Amtrak, began operation in 1971. Amtrak revenue-passenger miles have grown at an average annual rate of 2.9 percent from 1971 to 2006, rising to 5.4 billion passenger-miles in 2005. Commuter rail grew to about 9.5 billion passenger-miles in 2005, and rail transit passenger-miles grew to 16 billion in 2005 (Stacy C. Davis, Susan W. Diegel, and Robert G. Boundy, *Transportation Data Book: Edition 27,* ORNL-6081 [Oak Ridge: Oak Ridge National Laboratory, 2008], Tables 9.10–9.12). In total, these three rail transportation modes represented 30.9 billion

passenger-miles in 2005. At the same time, vehicle-miles per capita grew to 10,082 in 2005 (ibid., Table 8.2). This amounts to 3.2 trillion miles, based on a U.S. population of 296 million in 2005. Thus, the U.S. total passenger miles on Amtrak, commuter rail, and rail transit represent less than 1 percent of the total vehicle-miles traveled by U.S. passengers in 2005.

35. Energy Information Administration, *An Updated Annual Energy Outlook 2009,* Tables A2 and A17.
36. Daniel Sperling and Deborah Gordon, *Two Billion Cars: Driving Toward Sustainability* (New York: Oxford University Press, 2009).
37. Ibid.
38. S. Olz, R. Sims, and N. Kirchner, "Contributions of Renewables to Energy Security." International Energy Agency Information Paper (Paris: OECD, 2007).
39. World Bank, *Household Cookstoves, Environment, Health and Climate Change: A New Look at an Old Problem* (Washington, DC: World Bank, 2011).
40. E. Crewe, S. Sundar, and P. Young, *Building a Better Stove: The Sri Lanka Experience* (Colombo, 2010).
41. World Health Organization, 2005.
42. The World Bank, *2012 Data,* http://data.worldbank.org/indicator/EN.ATM. CO2E.KT?order=wbapi_data_value_2008+wbapi_data_value+wbapi_data_ value-last&sort=asc (accessed January 23, 2012).
43. See W. K., Biswas, P. Bryce, and M. Diesendorf, "Model for Empowering Rural Poor through Renewable Energy Technologies in Bangladesh," *Environmental Science and Policy* 4 (2001): 333–344; R. Krishnan, "Towards Energy Security: Challenges and Opportunities for India," paper presented to the Emerging Challenges to Energy Security in the Asia Pacific International Seminar, Center for Security Analysis, Chennai, India, March 16–17, 2009; K. R. Islam and R. R. Weil, "Land Use Effects on Soil Quality in a Tropical Forest Ecosystem of Bangladesh," *Agriculture, Ecosystems and Environment* 79 (2000): 9–16; D. H. Miah, Al Rashid, and M. Yong Shin, "Wood Fuel Use in the Traditional Cooking Stoves in the Rural Floodplain Areas of Bangladesh: A Socio-Environmental Perspective," *Biomass and Bioenergy* 33 (2009): 70–78; D. M. Kammen and M. R. Dove, "The Virtues of Mundane Science," *Environment* 39, no. 6 (1997): 10–41.
44. B. K. Sovacool, "Conceptualizing Urban Household Energy Use: Climbing the 'Energy Services Ladder,'" *Energy Policy* 39, no. 3 (2011): 1659–1668.
45. P. C. Stern and E. Aronson, *Energy Use: The Human Dimension* (New York: Freeman & Company, 1984).
46. R. Bent, L. Orr, and R. Baker, *Energy: Science, Policy, and the Pursuit of Sustainability* (Washington, DC: Island Press, 2002).
47. J. Gulliver and D. N. Zillman, "Contemporary United States Energy Regulation," in *Regulating Energy and Natural Resources,* ed. Barry Barton, Lila Barrera-Hernandez, and Alastair Lucas (Oxford: Oxford University Press, 2006), 113–136.
48. For example, over 60 percent of all export contributions to the budget of the Russian Federation come from sales of oil and gas. Keun-Wook Paik, *Sino-Russian Oil and Gas Cooperation: The Reality and Implications at 21* (2012).
49. B. K. Sovacool and M. Brown, "Competing Dimensions of Energy Security: An International Review," *Annual Review of Environment and Resources* 35 (2010): 77–108.

50. B. K. Sovacool, "Evaluating Energy Security in the Asia Pacific: Towards a More Comprehensive Approach," *Energy Policy* 39, no. 11 (2011): 7472–7479.
51. B. K. Sovacool et al., "Exploring Propositions about Perceptions of Energy Security: An International Survey," *Environmental Science and Policy* 16, no. 1 (2012): 44–64.
52. For example, non-hydrorenewable sources accounted for 3.7 percent or 752 billion KWh of globally produced electricity in 2010. See U.S. EIA, "International Energy Statistics."
53. Royal Dutch Shell, "Shell in Alaska News and Media Releases," http://www.shell.us/aboutshell/projects-locations/alaska/events-news.html (accessed May 25, 2013).
54. International Monetary Fund, "World Economic Outlook: Growth Resuming, Dangers Remain," *IMF Report 180* (April 2012).
55. Ibid.
56. Ibid.
57. S. Gordon, *Welfare, Justice, and Freedom* (New York: Columbia University Press, 1980).
58. John Rawls, *A Theory of Justice* (Cambridge: Belknap Press, 1971), 3–4.
59. William Antholis and Strobe Talbott, *Fast Forward: Ethics and Politics in the Age of Global Warming* (Washington, DC: Brookings Institution Press, 2010), 112.
60. See, e.g., J. Cao, "Reconciling Human Development and Climate Protection: Perspectives from Developing Countries on Post-2012 International Climate Change Policy," discussion paper 08-25, Harvard Project on International Climate Agreements, Belfer Center for Science and International Affairs, Harvard Kennedy School, 2008, http://belfercenter.ksg.harvard.edu/publication/18685/reconciling_human_development_and_climate_protection.html?breadcrumb=%2Fproject%2F56%2Fharvard_project_on_international_climate_agreements.
61. The UN Conference on Environment and Development, Rio Declaration on Environment and Development, Principle 7, http://www.unep.org/documents.multilingual/default.asp?documentid=78&articleid=1163.

2   Deciphering energy justice and injustice

Knowledge of what is does not open the door directly to what should be.
– Albert Einstein, Address to the Princeton Theological Seminary,
May 19, 1939

Introduction

In their assessment of the global energy system and its impacts on human health, an international research team of epidemiologists and environmental scientists recently reached a curious finding.[1] They noted that while some serious environmental burdens resulted from having too much energy – from waste, overconsumption, and pollution – others resulted from not having enough energy – from lack of access to modern forms of energy, under-consumption, and poverty. With increasing levels of wealth, these environmental burdens "shifted" in terms of severity and geographic and temporal reach (see Figure 2.1); thus, for instance, a decline in household environmental risks (through enhanced access to modern energy services, clean water, better healthcare, and the like) coincided with an increase in global risks such as climate change and other forms of transboundary environmental pollution. The team found that while solutions to some problems, such as indoor air pollution, obviously required an increase in wealth, solutions to other problems, such as climate change, might well require a diminution in wealth (at least wealth as it is conventionally defined). In other words, the current fossil fuel-based global energy system is a double-edged sword: It produces many benefits, but it also causes significant health burdens that shorten lives, undermine the conditions for happiness, and, in the words of the authors, "obstruct a more equitable and sustainable" future.

The findings of this research team should at the very minimum make it clear that any solutions to the problems associated with the energy system are not going to be simplistic. Proposals that advocate for "more of the same" in some form or another have not properly understood the complexity of the situation we find ourselves in – namely, that the global energy system is both a remarkable means for expanding opportunities and raising standards of living and an unwieldy mechanism that degrades health and enhances

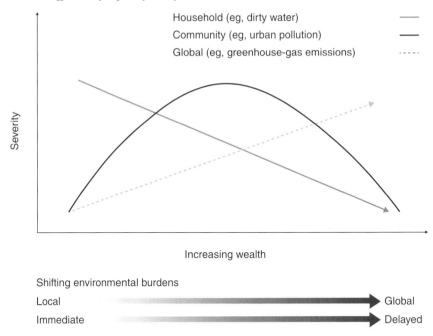

Figure 2.1 Transition of environmental burdens by scale, severity, and time

Source: Paul Wilkinson, Kirk Smith, Michael Joffe, and Andrew Haines, "A Global Perspective on Energy: Health Effects and Injustices," *Lancet* 370 (September 15, 2007): 965–977.

human misery. In many cases, these countervailing trends occur simultaneously. In this chapter, our point of departure is the indisputable fact that the global energy system, despite being a vehicle of economic and social development, conflicts with many of the demands of distributive, procedural, and cosmopolitan justice. For all intents and purposes, it operates without a moral compass.

In what follows, we draw from a range of modern philosophical theories in developing a framework for energy justice – focusing on Albert Einstein's "what should be" rather than merely "what is." We look at theories of distributive justice, which examine the so-called "outcomes" of justice (what is distributed and to whom and according to what principle of distribution), and theories of procedural justice, which look at the so-called "inputs" of justice (who participates in the decision-making process).[2] We also explore and borrow from aspects of cosmopolitan justice, with its emphasis on the global nature of the duties of justice.[3] Building on four core assumptions, we present a *prohibitive* principle of energy justice, which requires us to minimize harm through the design and operation of energy systems, and an *affirmative* principle of energy justice, which requires us to help those in need of energy services.

The energy system and social justice: An overview

What does it mean to speak of "energy justice"? According to a definition that has wide currency today, justice, or more specifically, social justice, is about the distribution of "benefits and burdens" in society.[4] So is energy justice concerned with how energy is distributed among members of society? Before we answer this, we must establish whether energy is a benefit or a burden. Most people, it is safe to assume, would answer that it is a benefit, or in more philosophical jargon, a good. We light our homes, offices, and streets with electricity; we cook our food and heat our buildings with natural gas; we power our cars and trucks with petroleum. Yet energy is a peculiar kind of good. Strictly speaking, it is not a good, but a means by which we acquire other goods. This understanding of energy is captured in a phrase used by energy analysts: "energy services." Energy does work for us – it provides us with services such as lighting, heating, and mobility. We are not interested in energy itself, but with the abundant benefits it provides.

Assuming for the moment that there is a good reason for making this distinction (which to the casual reader might appear a superfluous refinement), do we that conclude energy justice is about the distribution of energy services? In part, yes, but only in part – and not for obvious reasons. In fact, as this next section will elaborate, the distributive patterns of energy services within and across countries have implications for social justice primarily as an indirect consequence of the dominating nature of energy as a means.

Energy and distributive justice

Theories of distributive justice date back to the Greeks and are prominently associated with the work of modern political philosophers John Rawls and Ronald Dworkin.[5] Such theories concern themselves with how social goods are allocated across society. Distributive justice deals intently with three aspects of distribution: 1) What goods, such as wealth, power, respect, food, or clothing, are to be distributed? 2) Between what entities are they to be distributed (e.g., living or future generations, members of a political community, or all mankind)? and 3) What is the proper mode of distribution – is it based on need, merit, utility, entitlement, property rights, or something else? As we shall see, the distributive aspect of energy justice is, in part, about the distribution of energy services as a social good, but it is also, perhaps even more importantly, about how the harms of energy production and use are allocated.

Modern conventional energy systems, as we saw in Chapter 1, are designed around fossil fuels, which make solar energy that has been stored and concentrated over millions of years in solid, liquid, and gaseous forms available for human use. Coal, oil, and gas are easy to access and transport, and most importantly, by combusting them societies are able to obtain enormous amounts of work. Because the high-grade (low entropy) energy obtained from fossil fuels is a stock, whereas solar radiation is a flow, we can control the former to an extent impossible with the latter – especially with respect to the rate of use. As the economist Nicholas Georgescu-Roegen noted, "One generation, whatever it may do, cannot alter the share of solar radiation of any future generation."[6] When it comes to "mankind's entropic dowry," on the other hand, a few generations are capable of consuming the entire stock, as is becoming increasingly evident.[7] The result of this reliance on fossil fuels is that we have become accustomed to the daily employment of amounts of power that simply has no analog in the lives of people only a few generations ago. When quantities of energy of this order of magnitude are used to pursue traditional human ends, the ends are inevitably transformed in the process, sometimes beyond the point of recognition.

The use of energy today permeates every aspect of our culture and determines even our social relations. As Eugene Rosa and his coauthors observe:

> Energy, though fundamentally a physical variable, penetrates significantly into almost all facets of the social world. Life-styles, broad patterns of communication and interaction, collective activities, and key features of social structure and change are conditioned by the availability of energy, the technical means for converting energy into usable forms, and the ways energy is ultimately used.[8]

It is perhaps helpful to illustrate this point with an example. The purpose behind various forms of modern transport – fossil fuel-powered trains, cars, ships, and planes – is to move people and goods longer distances over shorter periods of time: The end result is greater mobility. Yet a moment's reflection should make clear that modern transportation has changed more than the way in which we get to our destinations: it has changed the destinations themselves. Our living spaces are now dominated by the means of transportation, so much so that we build cities around the needs of automobiles rather than the needs of men and women. The purpose and cultural significance of traveling has also changed, as has the meaning of staying at home. Indeed, the more attached we are to the use of modern forms of transport, the more uprooted and homeless we become.

As Ivan Illich pointed out 40 years ago, for those of us living in industrial countries, a large portion of our life is consumed one way or another by transportation: We spend hours a week stuck in traffic; a sizeable chunk of our annual earnings is used to pay monthly installments on the car, insurance, maintenance, parking, and fuel; we watch automobile commercials on TV; we spend enormous amounts of money in taxes on the health care system due to traffic accidents and the many illnesses that result from traffic-related air pollution. Illich notoriously calculated that the "model American puts in 1,600 hours to get 7,500 miles: less than five miles per hour," which is not much faster than the average person walks.[9] The defining characteristic of traffic in an industrialized world, according to Illich, is "more hours of compulsory consumption of high doses of energy, packaged and unequally distributed by the transportation industry."[10]

Energy is an instrumental good, and yet it has a tendency to dominate the ends to which it ought to be subordinate. The ends, indeed, are sometimes forgotten in pursuit of the means. The most fundamental reason to speak of "energy justice" – and thus bring together two seemingly incongruous concepts – is this: Energy makes available for human use historically unprecedented amounts of power, and this power, like all forms of power, harbors a potential for domination. The most basic form this domination takes is the instrumental domination of energy as a means over the ends it serves.

Modern sources of energy (lumps of coal, barrels of oil, cubic meters of natural gas, acid-leeched uranium) have become prerequisites today for the production and acquisition of a surprisingly large number of goods. The industrialized economy – which now reaches into almost every part of the globe – is entirely dependent on the energy services provided by modern energy systems: Manufactured goods, basic infrastructure, resource extraction, industrial agriculture, medicine, tourism, and international trade all require large inputs of energy. But so do many forms of social and political association, which rely heavily today on digital technology to provide public spaces that are otherwise lacking; education in many places has also become energy-intensive, and even less tangible cultural goods like social status depend to a large extent on energy consumption. For that matter, daily communication between neighbors and friends increasingly takes place through the media of electrically powered instruments. If we accept that social justice requires a rough equality of opportunity to acquire the basic goods of life and develop the capabilities that lead to a full and flourishing human life, then we have to count as critically important the distribution of energy services within and across societies.

Of course, the claim that we are entitled to energy services as a matter of justice makes no sense in isolation. In the early twentieth century, when much of the world had not yet been brought within the ambit of a modern industrialized economy, the absence of electricity in the highlands of Papua New Guinea was not a *cause célèbre* for advocates of social justice (and no doubt the highlanders would have looked unfavorably upon any efforts to supply them with such). But the world has changed a great deal since then. As we shall examine in the various chapters of this book, energy inequality and poverty in a globalized world exacerbate other forms of inequality: economic, social, and political. People who have no or limited access to energy services will generally have fewer educational opportunities, less access to fertile land and other natural resources, poorer health, negligible political representation, limited economic opportunities, and inadequate access to health services.

To the extent that energy justice *requires* an equitable distribution of energy services, this imperative must be understood as a consequence of the instrumental domination of energy, which has become, in many instances, the only means of obtaining basic goods to which human beings *are* entitled, like welfare, security, health care, education, and the right to employment. In other words, any entitlement to energy services is derivative of an entitlement to the basic goods that these services provide or make it possible to secure.

Energy and procedural justice

Another form of power imbalance that is a concern for energy justice is the economic and political domination exercised by social, industrial, and political elites who control the extraction and distribution of fossil fuels and electricity production. Because energy is central to the economy, it generates huge profits. Large revenues that are all but guaranteed provide a powerful incentive for companies and governments to manage carefully who has access to resources, how regulatory frameworks for extraction, transportation, and generation are structured, and which resources are available for exploitation. In the developed world, this sort of domination typically takes the form of "regulatory capture," campaign contributions, lobbying, and compliant politicians. In the developing world, as we shall see in Chapter 5, it can assume more violent and disruptive forms, including widespread corruption, social and political instability, human rights abuses, the resettlement of communities, armed conflict between factions, and even war. These issues all impinge upon procedural justice, which is about free prior informed consent for energy projects, representation in energy decision-making, and access to high quality information about energy.

The many injustices associated with the exploitation of energy resources – however endemic and brutal – are admittedly not intrinsic to energy as an organizational system; they are better described as instances of the garden-variety injustice of man toward man (or woman) that is a consistent and permanent feature of the human struggle for ascendancy. Nevertheless, given that the energy sector (as we define it in Chapter 1) is one of the dominant power structures within the world today, an adequate consideration of social justice must account for the role energy plays in distributing the benefits of economic and political power, and the social costs that result from these energy-related (re)distributive patterns. Likewise, an adequate assessment of the value of modern energy systems must acknowledge the enormous suffering and violence that is so often the "collateral damage" of profit-making from fossil fuels.

An energy justice concern that *is* more closely related to the inherent nature of energy systems is their tendency to promote a form of technocratic "authoritarianism," as we discuss in Chapter 5. Modern conventional energy systems as found in most industrialized countries are highly centralized and integrated, capital-intensive, and based on large-scale base load power plants and the deployment of sophisticated technology that requires specialized training to understand. The very scale of the North American electricity grid – called the largest and most complex machine in the world[11] – overwhelms the imagination. Undertakings of this nature require extensive collaboration of the industrial, administrative, financial, and political arms of the private and public sectors. They also require well-established lines of authority at various levels of government. The decision-making process is inevitably left in the hands of the experts in control, and democratic participation is minimal. As the historian of technology David Nye observes, "large-scale systems, such as the electric grid, do have some flexibility when being defined in their initial phases; however, as ownership, control, and technical specifications are established, they become more rigid and less responsive to social pressures."[12]

Admittedly, few people regard the energy system as authoritarian, and most would probably be surprised to hear it described as such. Perhaps this is because we perceive its purpose as serving the public good. Also, we are not aware of the extent to which public discourse about energy is directed by the interests of the energy industry. Public relations campaigns have been a feature of the electric industry from early in the twentieth century;[13] today, the coal industry and "big oil" spend hundreds of millions each year persuading the public of the importance of fossil fuels for domestic jobs and the national economy. Few people realize that there is nothing economically or technologically *necessary* about the energy system we have. While its current configuration is in part the unintended result of "path dependence" on earlier technology (see Chapter 7), it is also to a large extent the design of utility managers and their allies in finance and government who sought to "close the system" and "stifle radical inventions that could upset the central station paradigm and threaten established financial interests."[14]

Industry executives and politicians know that they can depend on a substantial segment of the population that simply accepts that the indirect environmental and health consequences of the energy sector (such as mountaintop removal coal mining, asthma and mercury poisoning, smog and climate change) are regrettable but unavoidable costs of a way of life we are not about to renounce. However, to recall the distinction we described at the beginning of this discussion, it is important not to confuse energy and energy services. Our attachment is to the latter. The question we need to ask is whether it is possible to provide the same energy services through a radically different energy system. The fact that there *are* technological alternatives to the conventional energy system, yet so few people (relatively speaking) have seriously questioned the reasonableness of the tradeoffs between the costs and benefits of the current paradigm, in itself indicates its authoritarian nature. Even in recent years, faced with a plethora of emerging alternative energy technologies that are commercially viable, widespread public concern about the environmental impacts of energy use, and a united scientific community calling for immediate changes to avert possible disaster, the authorities responsible for the energy sector still manage to present themselves as custodians and representatives of the only "realistic" alternative.

This aspect of conventional energy systems clearly runs afoul of most theories of procedural justice, which emphasize "public participation" and "due process" as constitutive elements of democratic governance.[15] A just society in procedural terms is one that legitimately claims the adherence of the governed, but such adherence can only spring from the individual's sense of being a willing and active member of a larger community. As Will Hutton writes, "membership of an associative conscience" requires "a social organization that allows every citizen the equal possibility of participation."[16] Or as geographers John Farrington and Conor Farrington put it: "A just society is one that *inter alia* grants the opportunity of participation in society to all of its members, and a society will certainly be unjust if it does not grant this opportunity to all of its members."[17] While many people today (though probably a shrinking majority) continue to accept the imposition of an energy system that affects everyone, for good and ill, but over which the average person has almost no control, it is an acceptance based on incomprehension and a lack of awareness of possibilities. Most authoritarian regimes that succeed do so because they are able to foster a culture of indifference, where questions are no longer asked because adequate answers are never given. Such regimes might be comfortable for some of us, but they hardly conform to the standards of democratic justice.

Energy and cosmopolitan justice

The last but probably most significant concern for energy justice is one we have already mentioned briefly: the global nature of the enormous environmental and human costs of the current energy paradigm. Recall that we started with David Miller's definition of social justice, that it concerns the distribution of benefits and burdens in society. We confirmed that energy justice is in part about the distribution of energy services as a social good. But as the preceding discussion should have made clear, any concern for energy justice, at least within the context of the conventional energy system, will inevitably end up focusing much more closely on the distribution of burdens or "bads": the environmental and social costs that indirectly result from extracting, producing and using energy, but are not included in the market price. Another way of putting this is to say that a book about energy justice will end up being a book about energy injustice, as will become clear starting in Chapter 3, which looks at the environmental harms arising from the energy sector.

Many of the problems of air pollution associated with power plants are transboundary in nature; in the case of climate change, they are global. Our approach to energy justice therefore shares many of the concerns and difficulties of cosmopolitan justice. For instance, how should the community of justice be defined? Is it possible to speak of universal or cross-cultural human needs? Theorists of cosmopolitan justice argue that the principles of justice – such as those from distributive and procedural justice – must apply universally to all human beings in all nations. Cosmopolitanism implies that "duties of justice are global in scope, and these duties require adherence to general principles including respect for civil and democratic rights and substantial socioeconomic egalitarianism."[18] Put another way, cosmopolitan justice accepts that all human beings have equal moral worth and that our responsibilities to others do not stop at borders.[19] It should be clear that the transboundary nature of energy injustice requires a similar conception of the reach of moral and political responsibility.

But if fundamental principles of justice should not be confined by spatial considerations, what about temporal considerations? Do human needs and aspirations remain the same across time? The potentially long-term impacts of climate change, and the even longer-term impacts of radioactive waste from uranium mining and nuclear power plants, mean that any attempt to think about the implications of energy justice also has to address the question of intergenerational justice: How it is possible to bring within the community of justice human beings that do not yet exist? A parochial understanding of energy justice, in short, is not compatible with the far-reaching nature of many of the impacts of the energy sector, which extend outwards in space and time and before long will embrace all of humanity.

We shall address these matters of universality in the second part of this chapter. Before turning to it, however, it is important to note a significant qualification to our understanding of energy justice: it is not focused on environmental harms *per se*. To the extent that energy justice addresses the ecological damage caused by energy systems, it is concerned with the ramifications of this degradation for human well-being. In this respect, energy justice is similar to the environmental justice movement, and deals with some of the same issues. For instance, in Chapters 3, 6, and 7 we touch upon the differential social, economic and geographic impacts of energy externalities such as air pollution, mine tailings, energy-related accidents, and climate change. Because of this focus on human welfare, we do not have to enter the contentious debate over whether it makes any sense to speak of justice in relation to non-human species and the natural world.[20] This is an important question, of course, and obviously has some bearing on how we use energy; but it deals with a much wider range of concerns, and involves considerations that are not necessary to resolve the fundamental issues of energy justice. The claim that human beings have a moral responsibility to the non-human world certainly lends weight to the concerns of energy justice, but dismissing this assertion (at least in this book) in no way relieves us of the imperative to consider carefully how our energy systems affect the well-being of human beings, now and in the future.

Philosophical framework for energy justice

The philosophical framework outlined in the remainder of this chapter begins with four assumptions. From these assumptions, we arrive at two principles of energy justice. Our intention is that these principles will serve as guides for thinking, and acting, on energy-related questions. They are sufficiently comprehensive to cover most issues that arise within the contexts of developing and implementing energy systems. At the core of our conception of energy justice is recognition of the imperative to respect the dignity of each and every human being. In the words of Article 1 of the Universal Declaration of Human Rights, we affirm that "[a]ll human beings are born free and equal in dignity and rights."[21]

In what follows, it is not our intention to ground this affirmation in a foundational theory of justice, and we deliberately avoid when possible taking a position on foundational questions. We take it for granted that the values of equality and human rights are firmly embedded in the attitudes, beliefs and expectations of the vast majority of people today; even those who disagree with them will usually find it necessary to frame their disagreement in language that is borrowed from the discourse of equal rights; and those who violate these values will still find it expedient to pay lip service to them.

The reason we remain uncommitted with respect to any foundational theory of justice is twofold. First, as authors we do not necessarily agree ourselves on all foundational questions, so it was better to avoid them. As we discovered, however, disagreement at the level of theory did not get in the way of agreement at the level of practice. Our own experience thus confirmed a working assumption that the two principles of energy justice at the heart of our approach should be able to accommodate a wide range of philosophical theories, cultural beliefs and political positions. This, indeed, is the second reason for remaining uncommitted on foundational questions. The import of this book is ultimately practical. Our aim is not to have readers ponder the foundations of justice. Rather, we hope to influence energy analysts, business leaders, politicians, utility managers and energy consumers to consider the implications for social justice of every aspect of our current energy system.

Assumption 1: Every human being is entitled to the minimum of basic goods of life that is still consistent with respect for human dignity

In her remarkable study of the basic requirements of a just society, *The Need for Roots,* the French philosopher Simone Weil noted that the ancient Egyptians "believed that no soul could justify itself after death unless it could say: 'I have never let anyone suffer from hunger.'"[22] According to Weil, the one obligation human beings have to one another is respect. But this obligation "is only performed if the respect is effectively expressed in a real, not a fictitious, way; and this can only be done through the medium of Man's earthly needs." The notion of respect for human dignity is closely associated in the modern period with Immanuel Kant. For Kant, the dignity of the person consists in being an autonomous rational agent; the moral law is based on respect for human rationality, which requires that we treat every individual as an end and not a means. The most famous modern philosopher of justice, John Rawls, follows Kant (implicitly) in also making personhood reside in (moral and prudential) rationality.[23] We do not take this approach in our understanding of what constitutes human dignity, but reach further back in the tradition to Aristotle's notion of the human being as a "political animal." In this respect, we follow closely the account of dignity articulated by the American philosopher Martha Nussbaum in her more recent writings.

Nussbaum points out that the "animal dignity" peculiar to human beings includes not only rationality but also sociability. Moreover, "bodily need, including the need for care, is a feature of our rationality and our sociability; it is one aspect of our dignity, then, rather than something to be contrasted with it."[24] "We have a claim to support in the dignity of our human need itself."[25] Nussbaum explicitly identifies herself within the Aristotelian/Marxian tradition of regarding the human being as a being with "rich human need" – to use Marx's phrase.[26] She writes that "need and capacity, rationality and animality, are thoroughly interwoven, and that the dignity of the human being is the dignity of a needy enmattered being."[27] Needs and capacities are both sources of moral claims; we respect the dignity of other human beings by responding to the moral claim their needs make upon us.

The most basic needs of human beings are for the goods of security and welfare, or what Brian Barry calls "vital interests": adequate nutrition, clothing and shelter, bodily integrity, clean drinking-water, health care, education, and so forth.[28] One characteristic of such goods is that they are necessary regardless of what a person values in life – in the phrase of Andrew Dobson, they are "preconditional goods."[29] As Dobson points out, a subset of preconditional goods (and one that is important for the concerns motivating the present inquiry) is those environmental goods (for example, unpolluted water or the maintenance of the ozone layer) that are essential for human existence. (We can leave aside for the moment the consideration that some environmental goods are "public goods," like clean air, that cannot be distributed strictly speaking. In this case, what society distributes are the rights to use or damage such public goods.) A society with any claim to moral legitimacy will provide for the equitable distribution of these basic goods. Edmund Burke made the point concisely more than two centuries ago: "Government is a contrivance of human wisdom to provide for human wants. Men have a right that these wants should be provided for by this wisdom."[30] More recently, the concept of need has come to occupy a central position within the discussion of sustainable development. Thus the influential 1987 Brundtland Report, *Our Common Future,* defined sustainable development as "meeting the needs of the present without compromising the ability of future generations to meet their own needs."[31]

As to how such goods should be distributed, in what proportion and to whom, that should be obvious: in accordance with the need of the recipient. Michael Walzer defended the idea that there are numerous "spheres of justice," each with its particular sort of good, and that the principle of distribution for each sphere was determined by the good in question. In the case of "security and welfare," the distributive principle clearly is need. Walzer describes the principles governing the sphere of security and welfare in the following terms:

> that every political community must attend to the needs of its members as they collectively understand those needs; that the goods that

are distributed must be distributed in proportion to need; and that the distribution must recognize and uphold the underlying quality of membership.[32]

A pair of theoretical and practical problems arises, however, when we apply these concepts to the concerns of energy justice. As we have already remarked, the environmental impacts of the energy sector often extend beyond the boundaries of nation states; in the case of greenhouse gas emissions, the impact is global. As a consequence, energy justice is also global in character and raises the question of international justice. But applying the concept of need as a distributive principle of justice across different communities and cultures is more problematic than it might at first appear.

The problem is referenced by Walzer when he speaks of needs in terms of the "collective understanding" the political community has of them. What counts as a need is something that the community decides for itself.[33] As Walzer explicitly denies the possibility of a community that encompasses "all men and women everywhere," this suggests that it is a mistake to speak of human needs that are universal, as there is no corresponding community with its "world of common meanings" to define and identify them as such.[34] "The nature of a need is not self-evident," writes Walzer.[35] In contrast to the cosmopolitanism of Kant, which begins with the individual rational agent and the rights that belong to rationality, Walzer takes the person embedded within a particular culture with its own history and frame of reference as the starting point. It is difficult not to countenance Walzer's claim that the "category of socially recognized needs is open-ended."[36] Is primary school education, for instance, a "basic need" in every culture? Not if we mean by it a particular kind of education, say, that which is typical in most Western nations today. But this argument only goes so far. Education as a form of cultural initiation and preparation *is* a basic need in every country and culture; presumably the most suitable kind of education will be that which best prepares the student for the world he or she is entering. Today, for the vast majority of people, that world is one deeply influenced and shaped by the currents of economic globalization, and this fact needs to be accounted for in the type of education that is provided.

With respect to those goods that we are most concerned with in the context of energy justice, the preconditional goods jeopardized for countless thousands of people by the global energy system – clean drinking water, the secure possession of land, the right to be safe from bodily harm – it makes little sense to say that such "vital interests" are not universal human goods. (We can reject out of hand Walzer's claim that even the need for food is in some way culturally conditioned.[37]) Walzer seems to be minimizing the significance of the animality of human beings over against cultural situatedness, just as Kant minimized it over against human rationality. Surprisingly, Kant himself hints at a different way of approaching cosmopolitan justice than one based on either individual rationality or cultural parochialism – an approach that is more in keeping with our own preoccupations – when he writes that "the right to the earth's surface which belongs to the human race in common could finally bring the human race ever closer to a cosmopolitan constitution."[38]

In developing her own theory of cosmopolitan justice, Nussbaum draws attention to the importance of another consideration that is not adequately accounted for within the parameters of Walzer's reflections on community. "[T]he duty not to use people as a means," she writes, "cannot be plausibly separated from critical scrutiny of the global economy and its workings, and thus from a consideration of possible global redistribution and other associated social and economic entitlements."[39] To deny the claims of international justice based on the lack of a genuine global community is to ignore the reality of the international economic order and what Amartya Sen calls the world's "interdependence of interests," in which the actions of a multitude of different commercial, industrial, political, and cultural units all affect one another in myriad ways.[40] Undeniably, there is at the very least a "community of interests" among diverse people and countries today, and this is or ought to be an adequate basis upon which to erect some form of international justice.

The attempt to conceptualize an international order of justice based on the distributive principle of need has to contend with another related objection, however. David Miller, like Walzer, takes a contextual approach to justice. He also argues that the principle of need cannot operate across nation states, because it is only within communities (such as nation states) that there is the necessary reciprocity to establish the involuntary and cooperative relationships that make possible the recognition of need as a claim of justice.[41] Between states, the distribution of goods to satisfy needs is a matter of humanity as opposed to one of justice. The philosopher Thomas Nagel likewise argues that without the necessary institutional mechanisms at the international level (which today are clearly lacking, notwithstanding the United Nations) it makes no sense to speak of global justice. As he notes, "It seems to me very difficult to resist Hobbes's claim about the relation between justice and sovereignty," and "[i]f Hobbes is right, the idea of global justice without a world government is a chimera."[42] For this reason, Nagel emphasizes the importance of the demands of "minimal humanitarian morality" in the global context.[43]

The distinction between justice and benevolence is an ancient one in political thought. Cicero spoke of the line between giving people what they are entitled to as opposed to what we give them out of generosity without acknowledging any claim.[44] It is a distinction that is very much a part of the international climate change negotiations, where there is a clear reluctance on the part of polluting nations to recognize a level of international responsibility that would *require* them (as a matter of justice) to help poor and more vulnerable countries adapt to the impacts of climate change; financial contributions, according to polluting nations, should be on a voluntary basis (as a matter of humanitarianism).

In the case of climate change, indeed, there is a strong argument to be made that the mutual interdependence of countries in tackling the global problem of reducing greenhouse gas emissions actually establishes the requisite condition of reciprocity for the recognition of need. Yet the objection still carries some weight. Intuitively, most people would probably think it is humanity rather than justice that prompts the international community to respond, say, to the aftermath of an earthquake in Japan. On the other hand, as Amartya Sen convincingly points out, "When people across the world agitate for *more* global justice . . . they are not clamoring for some kind of 'minimal humanitarianism.' Nor are they agitating for a 'perfectly just' world society, but merely for the elimination of some outrageously unjust arrangements to enhance global justice . . . on which agreements can be generated through public discussion, despite a continuing divergence of views on other matters."[45]

The celebrated philosopher John Rawls recognizes this problem in his own discussion of international justice, which contemplates extending the principles of the social contract to the relations between nation states. While it is tempting to see his "difference principle" – that inequalities among members of society should only be allowed to the extent that they benefit the worst off – as providing a good framework for thinking about the problems of energy justice, its usefulness is diminished by the fact that one of the essential conditions for "justice as fairness" is a rough equality of power and capacity among contracting members of the just society. This condition is clearly not met in the international context, where there are gross disparities of power and influence among nation states. Rawls defines society as "a cooperative venture for mutual advantage."[46] What situation of possible mutual advantage could ever exist between the United States and the island nation of Tuvalu, which is slowly being submerged by rising sea levels, in part due to energy consumption in the United States?

In a later work, *The Law of Peoples*, Rawls acknowledges that his principles of justice as fairness cannot be properly applied in the international context, and concentrates instead on a more limited form of cooperation, wherein countries negotiate about general principles of civility and humanitarian behavior.[47] But this is clearly not a satisfactory way of articulating a basis for the manifest *duties of justice* that a country such as the United States owes to a country such as Tuvalu. The citizens of a nation who are losing their country itself as a result of the use of energy in other nations are not appealing to the *benevolence* of others by requesting that action be taken to halt the increase in concentration of atmospheric carbon dioxide – they are simply (and quite rightly) demanding *justice*.

While we do not wish to minimize the very real practical and theoretical difficulties associated with the notion of international justice, the proper setting for energy justice (in every respect) is the entire globe, not individual countries. We will address this problem again when we look at the principles of energy justice. In general, however, we follow the "cosmopolitan" approach to these questions articulated in the works of Amartya Sen and Martha Nussbaum.

Assumption 2: The basic goods to which every person is entitled include the opportunity to develop the characteristically human capacities needed for a flourishing human life

Is the distributive principle of need an adequate basis for addressing the various concerns of energy justice? Returning to our starting point, we might also ask whether respect for human dignity as required by justice is satisfied simply by ensuring that human needs are met. Amartya Sen observes in this context: "Certainly, people do have needs, but they also have values and, in particular, cherish their ability to reason, appraise, choose, participate and act. Seeing people only in terms of their needs may give us a rather meager view of humanity."[48] Of course, the answers to these questions partly depend on how need is defined. As we have seen, Nussbaum's understanding of the concept of "rich human need" encompasses the development of the specifically human faculties of reasoning, choosing, judging, valuing, and so forth. According to Nussbaum, participation in the life of a community is equally a human need, alongside the more basic needs for food and warmth.

Nussbaum and Sen have both developed theories of justice centered on the core idea of human capabilities.[49] In Nussbaum's case, this focus can be seen as a deliberate development of the Aristotelian tradition. As contrasted with a rights-based approach, Nussbaum writes that the capabilities approach:

> understands the securing of a right as an affirmative task. This understanding has been central to both Sen's and my version of the approach. The right to political participation, the right to the free exercise of religion, the right of free speech – these and others are all best thought of as secured to people only when the relevant capabilities to function are present. In other words, to secure a right to citizens in these areas is to put them in a position of capability to function in that area.

The emphasis on capabilities appeals to an Aristotelian account of the goods of human life. For Aristotle, it is important to distinguish between the *means* of a flourishing human life and the *ends* to which these means are put. Moreover, the primary difference between mere life and the "good life" is being able to share actively in the benefits of political association.[50] As philosopher John O'Neill usefully summarizes Aristotle: "Given the kinds of beings we are, friends, family and fellow citizens are goods, and a person without them could not live a flourishing life. Second, these relationships make accessible to us a variety of goods that could not be realized alone or within smaller associations."[51] The ideal of what Sen calls "participatory living" is at the heart of Aristotle's understanding of human nature.[52]

For Sen, the significance of the capabilities approach is its emphasis on "the lives that people can actually live" and "the freedoms that we actually have to choose between different kinds of lives."[53] Justice is not simply a matter of devising appropriate institutions and social arrangements; it should never lose sight of the actual opportunities that people have to develop and enjoy their own capabilities. As Sen points out, this approach to justice draws attention to the responsibility we all have for what we do and become. "Since a capability is the power to do something, the accountability that emanates from that ability – that power – is a part of the capability perspective, and this can make room for demands of duty – what can broadly be called deontological demands."[54]

The emphasis Sen and Nussbaum put on capabilities coincides with the concerns of energy justice in two important respects. First, when Nussbaum writes that securing a right to citizens in a given area means putting them in a position of capability to function in that area, it should be clear that in many cases, especially in the developing world, this will mean providing them with the energy services needed to function adequately. With its focus on the conditions that make possible the exercise of a right, rather than the mere granting of the right, the capability-centered perspective provides a useful framework for thinking about the role of energy services in helping or hindering the development of human capacities.

Another reason to consider the capabilities approach is that it offers an analytical tool for examining the paradoxical situation most of us find ourselves in with regard to modern energy systems. On the one hand, the use of energy puts into our hands the ability to increase the reach of our activity and the range of our experience – it empowers us as individuals. On the other hand, it should be evident that the vast majority of people have very little say in how energy is provided, where energy resources are extracted, what sorts of resources are developed, to whom energy is provided, who bears the costs of energy extraction and production and who enjoys the benefits, and most importantly, how decisions on all these matters are reached. We live in democratic societies attached to the idea of some sort of participatory governance, yet the energy systems that are central to our way of life both as individuals and communities are, for the most part, completely beyond our control. Energy, which gives us power as individuals to accomplish so many things, at the same time renders us powerless as citizens, unable to exercise any control over decisions which have profound and increasingly negative impacts on our lives.

When confronted with the energy system, dominated by huge corporations and complex bureaucracies, most of us fall back on a passive acceptance of the status quo. And yet the consequences of this passivity about so crucial a matter, which leaves decision-making authority largely in the hands of vested interests and power-hungry governments, are becoming difficult to ignore. Increasing global ecological degradation, numerous military conflicts about access to energy resources, political corruption fuelled by profits from the energy sector, growing social marginalization as energy consumption exaggerates social and economic inequalities, and mounting concern over long-term geopolitical stability as energy security becomes a more urgent concern for nation states – foci for the rest of this book – all point to the need for fundamental changes in the energy system.

The capabilities approach to justice drives home the fact that our civil and political rights are impotent without effective participation in the decisions that shape energy systems. Further, it emphasizes the fact that the securing of these rights is in Martha Nussbaum's phrase "an affirmative task," not something that can be assumed or simply taken. Finally, it underlines the responsibility we all bear for the final outcome of decisions made regarding the present and future development of global energy systems, both as individuals and political communities. Whether or not we encourage and ensure the design and implementation of energy systems that are more responsive to the needs of people is ultimately a choice about how we want to live and what capabilities we want to exercise.

At a more practical level, the actual business of decision-making is a central focus of procedural theories of justice, which are oriented toward process: the fairness and transparency of decisions, the adequacy of legal protections, and the legitimacy and inclusivity of institutions involved in decision making.[55] Procedural justice deals with recognition (who is recognized); participation (who gets to participate); and power (how is power distributed in decision-making forums).[56] It might be helpful at this point to give an example of an already established practice that incorporates the kinds of considerations we have in mind, and ensures that there is a genuine opportunity to "participate effectively in political choices that govern one's life."[57]

The protections outlined by procedural justice are perhaps best enshrined in the idea of free, prior, and informed consent. Free, prior, and informed consent refers to "a consultative process whereby a potentially affected community engages in an open and informed dialogue with individuals or other persons interested in pursuing activities in the area or areas occupied or traditionally used by the affected community."[58] Its main characteristics are that it is "freely given," "obtained before permission is granted," "fully informed," and "consensual."[59] "Freely given" implies that no coercion, intimidation, or manipulation has occurred: that potentially affected people freely offer their consent. "Prior" implies that consent has been sought sufficiently in advance of any meaningful decision to proceed with a project, before things like financing or impact assessments begin. "Fully informed" means that information about the project is provided that covers its nature, size, pace, reversibility, and scope; expected costs and benefits; the locality of areas to be affected; personnel and revenues likely to be involved; and procedures for resolving conflicts, should they occur. In other words, communities affected by energy projects must understand their rights, and the true implications of the projects as well as they can be known, so that opponents and proponents can negotiate with equality of information. Lastly, "consent" means "harmonious, voluntary agreement with the measures designed to make the proposed project acceptable."[60] It does not necessarily mean complete consensus; it is, however, distinct from consultation – the act of merely discussing a project with a community – because it gives communities the ability to "say no."[61]

In theory, the implementation of free, prior, and informed consent means that any type of energy project that depends on the use of force, requires involuntary action, or contributes to poverty or social instability is unacceptable. It makes it possible for communities to refuse projects and also ensures that they cannot be resettled involuntarily – that resettlement packages are attractive and that communities consent willingly. As the United Nations Commission on Human Rights recently explained it, "free, prior, and informed consent recognizes indigenous peoples' inherent and prior rights to their lands and resources and respects their legitimate authority to require that third parties enter into an equal and respectful relationship with them, based on the principle of informed consent."[62]

If we start from the assumption that human capabilities, including the right to political participation, are basic goods to which we are entitled by justice, it is clear that energy justice implies meaningful involvement and access to the decision-making process for energy systems. A just procedure would ensure the availability of adequate information about a given energy project, and opportunities for genuine participation and informed consent. It would seek to include and represent minorities and all stakeholders at every stage of the project's decision-making process, from agenda setting and formulation to siting and evaluation. It would also provide access to legal processes for challenging violations of energy rights.

Assumption 3: Energy is only an instrumental good – it is not an end in itself

Our third assumption might seem blatantly obvious, yet one could be forgiven for thinking otherwise given the history of energy policy over the past three-quarters of a century. The unquestioned belief motivating past and to a great extent present energy planning has been that more energy in itself is a good thing: maximization of consumption – with energy, as with everything else – has been the operative policy.[64]

Probably the most compelling reason for this policy focus has been the assumption that growth in energy consumption is strongly correlated with an increase in economic growth. Sociologists Eugene A. Rosa, Gary E. Machlis, and Kenneth M. Keating argue that from the 1940s to the early 1970s energy policies in the United States and elsewhere were dominated by economists who emphasized the importance of energy consumption to the economic performance of societies.[65] Studies compared indices such as Gross National Product with the amount of energy a given country consumed. Energy policy, therefore, was directed toward devising strategies to ensure a supply of energy that was adequate to guarantee economic growth. The energy crisis of the 1970s and the publication of studies by Amory Lovins and others helped to upset this complacent assumption.[66] These new studies showed that industrial economies differed in their energy consumption, and that beyond a threshold level necessary to achieve industrialization, wide latitude in the amount of energy needed to sustain living standards existed, as can be demonstrated easily today if one compares the energy consumption in Europe with that in North America. Yet this questioning of the dominant model of energy policy has not really displaced the assumption that economies of scale in energy production can still be realized. The fundamental market-based belief in a link between GDP and energy consumption, therefore, continues to dominate thinking about energy policy in many parts of the world.[67]

Once we discard this questionable belief, however, and start with the assumption that energy is only an instrumental good, the first consequence is that it becomes more important to ask what the ends are for which energy is the means. In other words, the demand for more energy is meaningless in itself, because energy is only a mechanism for providing certain goods, namely, energy services. But energy services can be provided in many different ways and on many different scales. Most importantly for our present concerns, *how* they are provided cannot be separated from an analysis of the kinds of social and economic structures they both reflect and reinforce, along with the distributive patterns of benefits and burdens they engender.

This point is reflected in a list of four maxims for the analysis of energy policy proposed by one of this book's authors and the Nobel Laureate Marilyn A. Brown. The second maxim, symmetry, holds that the study of energy technologies must focus on both social and technical issues:

> Focusing on the symmetrical dimensions of technological development has at least two implications for understanding the evolution of energy technologies. For one, it reminds us that the current energy system – with its gas stations, oil refineries, electric substations, transmission lines, expansive natural gas pipelines, coal mines, and varying types of generating and consuming technology – was and is by no means inevitable. Instead, each of these technologies is the product of social negotiation and compromise. Since the current system was chosen and elaborated upon by actors, it can also be changed by human participants. Additionally, making visible the contingency of the energy technologies allows us to study and analyze the factors that make current technologies socially acceptable. In other words, symmetrical analysis helps show us what social conditions are necessary for a given technology (or set of technologies) to succeed, at the same time such conditions may make other technologies unacceptable.[68]

Analysis of the benefits and costs of any proposal for extending, modifying, supplementing, or replacing the current energy system, in other words, requires a detailed examination of a large number of issues, on top of those typically investigated. What alternatives are available for providing the same energy services without increasing energy production? How can the required energy services be most equitably provided? For whose benefit is the proposed project, and why is it being promoted? What will the direct and indirect impacts of the proposed project be on existing economic and social structures? What is the appropriate scale of energy required for meeting the postulated need for energy services, and are the benefits localized, or distributed more extensively? Who will bear the greatest share of whatever burdens are associated with the proposed project, and are they in a position to consent to these burdens? Questioning along these lines accommodates the ethical dimension of energy policy; moreover, it acknowledges that a just energy policy requires that such considerations be given equal weight to technical and financial constraints and opportunities.

Once we begin challenging how energy technologies fit into and around existing social and economic structures and institutions, it soon becomes clear that under the current energy paradigm it is usually the social and economic structures that have to accommodate themselves to the various mechanisms for providing energy – a phenomenon we have already remarked on when discussing the "instrumental domination" of energy. The last and most important maxim proposed by Sovacool and Brown is one of energy prudence, which borrows from the work of Lewis Mumford and E. F. Schumacher in promoting the idea of "appropriate technology."[69] This concept has much in common with Mumford's notion of democratic technology, which he describes as person-centered, resourceful, and durable, drawing upon small-scale methods of production, local ownership, and control, and seeking to conform as much as possible to the natural environment.[70] E. F. Schumacher also argued against the domination of large and capital-intensive technologies, especially in the developing world where supporting institutional structures were usually absent, pointing out that projects based on such technologies often failed to deliver the promised benefits, and brought about unforeseen and unfortunate consequences, because they were implemented on the wrong scale. The concept of scale, indeed, is implicit in his concept of "intermediate technology," which later developed into "appropriate technology." According to Schumacher, intermediate technologies must create jobs where people live, rather than requiring them to migrate to urban centers; they must be cheap enough to function without too great a need for outside sources of capital; they must utilize the simplest production methods available, to minimize the need for non-local skills and knowledge; and, when possible, they must integrate local materials for local use.[71]

This is not to say that an energy policy responsive to the concerns of justice will have no place for centralized, large-scale energy systems based on capital-intensive and sophisticated technology. But it will be more sensitive to the circumstances – understood comprehensively – within which such systems operate. We have yet to overcome completely our bewitchment with the power offered to us by modern energy systems – not surprisingly. But we can no longer afford to remain bewitched. An appreciation for the merely instrumental and subsidiary role of energy is a necessary first step in the development of a sound energy policy that considers the entire range of possible impacts that a given energy system can have on society, including its social, cultural, economic, and environmental consequences. Most importantly, as far as energy justice is concerned, the use of energy must be determined by the human ends it serves (rather than these ends being distorted to fit the technical imperatives of energy), and these ends must be consistent with respect for the equal dignity of human beings.

Assumption 4: Energy is a material prerequisite for many of the basic goods to which people are entitled

As we have already discussed at some length, modern societies are entirely dependent on the use of energy. Abundant energy sources form the foundation upon which the industrialized economy is built. When a blackout occurs like that which blanketed parts of the United States and Canada in 2003, economic activity virtually ceases. This basic fact about modern society – its complete dependence on energy – is not simply a concern for engineers and technicians, however, but has profound implications for energy justice.

Discussing the relationship between political and civil liberties and social and economic rights, Nussbaum observes that even in a developed society freedom of speech and political freedoms have "material prerequisites."[72] In other words, it makes no sense to guarantee liberties that people cannot possibly actualize due to economic or educational deprivation. A theory of justice that rejects the need for any kind of economic redistribution or provision of communal goods cannot plausibly argue that it nevertheless defends the priority of civil rights and liberties. Securing adequate nutrition, shelter, education, and employment are necessary conditions or "prerequisites" for the realization of civil and political freedoms. Thus Nussbaum cautions that "[t]he capabilities approach insists throughout on the material aspects of all the human goods, by directing our attention to what people are actually able to do and to be." The approach "stresses the interdependency of liberties and economic arrangements."[73]

We take this analysis one step further. Because of the near-total dependence of modern economic and social systems on energy, energy has become a material prerequisite for many of the basic goods of human life. The importance of this assumption relates to the question of whether it makes sense to speak of having a right to energy, or more accurately, to energy services. Human beings do not have a fundamental right to energy services, but they do have a fundamental right to certain goods that today are difficult if not impossible to provide without energy services. To give a prosaic example, adequate shelter falls within those basic goods we described under the heading of security and welfare, the need for which is recognized as making a claim on justice. But in many cities today, the only means of heating a house or apartment, and thus providing "adequate" shelter, is through access to natural gas or some other form of energy. Without the material prerequisite of the energy service, it is impossible in this case to meet the need for adequate shelter. As we shall see in a later chapter, this situation in fact describes the reality of many people today who struggle with fuel poverty, with consequences ranging from increased risk of respiratory and circulatory illness in adults to thousands of excess winter deaths among children and the elderly.

Energy services, however, are a material prerequisite for more than just adequate housing and mobility. This is certainly the case in the developed world, but with globalization and the integration of all countries into international markets, it is increasingly the case in the developing world as well. A person's ability to benefit from such diverse goods as education, employment, communication, effective participation in the community, social standing, and health services is affected to varying degrees, depending on circumstances, by her level of direct or indirect access to energy services. Owning a computer, for instance, is a great advantage when it comes to education and employment prospects; having access to one has become almost a necessity for any but the most menial of occupations. As we shall see, a lack of access to energy services is one of the greatest impediments facing many people throughout the world today. This becomes clear when we direct our attention "to what people are actually able to do and to be."

From assumptions to principles

From these four assumptions, we develop two principles of energy justice. As already stated, we offer these principles as practical guides for decision making. In other words, we do not claim that they represent the last word on the ethical implications of energy systems, or cover all eventualities that might possibly arise with respect to energy and the demands of justice. But we do believe that they are comprehensive enough to encompass the majority of issues that arise in the course of developing and implementing energy systems. Moreover, they have the advantage of simplicity, which is a virtue when it comes to making sense of the conflicting interests, constraints, expectations, and ramifications that characterize the energy sector.

The prohibitive principle

The Prohibitive Principle states that "energy systems must be designed and constructed in such a way that they do not unduly interfere with the ability of any person to acquire those basic goods to which he or she is justly entitled."

As its name implies, this principle "prohibits," it sets a constraint on human actions so that the damage inflicted on people by energy systems is minimized. It asks us simply to refrain from doing what we know will cause existing or future generations harm. All energy systems have their associated costs:

Mining for coal and uranium produces toxic waste; power plants pollute the air with particulate matter, sulfur dioxide, mercury, and carbon dioxide; wind farms clutter the landscape; hydroelectric dams flood valleys, displace communities, and degrade aquatic and riparian habitats. What is striking about these costs, however, is that they do not affect all people equally. The burden of environmental risk is in most cases borne by those least able to protect themselves against it. This disproportionate impact of energy systems on the poor and vulnerable is one of the principle themes of the environmental justice movement. As sociologist Andrew Szasz writes in his history of the movement: "Toxic victims are, typically, poor or working people of modest means. Their environmental problems are inseparable from their economic condition. People are more likely to live near polluted industrial sites if they live in financially strapped communities."[74]

The impacts of global climate change likewise affect the poor and marginalized of the world to a disproportionate degree, in part because they are more exposed and vulnerable to extreme weather events, in part because they do not have the financial or institutional resources to adapt to the changing climate (in academic parlance they lack *adaptive capacity* and *resilience*). Again, looking at another aspect of the global energy system, the many conflicts that arise due to competition for access to energy resources typically affect people who in no way benefit from the wealth generated from these resources, but nevertheless find themselves uprooted from their homes, marginalized, harried and tortured, and in the worst cases, embroiled in vicious military or paramilitary conflicts.

The prohibitive principle addresses the many environmental and social costs or negative externalities of the energy sector. Starting from the recognition that justice entitles every human being to certain basic goods – including the preconditional goods of clean water and air, access to food and shelter, and so forth, and also the goods that can only be realized through participation in communal life and the development of capabilities – it requires those involved in the production and generation of energy systems to minimize to the greatest degree possible the negative consequences these systems have for others.

It is significant that the prohibitive principle speaks of "undue interference." As already stated, few energy systems are cost-free. The question is whether there is an equitable sharing of the burdens and benefits of the energy system. The great virtue of Rawls's "difference principle" is that it recognizes that economic and social inequalities can in many circumstances improve the situation of those worst off – defined in terms of means of access to a self-respecting existence – but that these inequalities are only permissible to the extent that they are to the greatest benefit of the least advantaged members of society. We have already mentioned the reasons why this principle is problematic in the context of international justice; nevertheless, the basic idea behind it is incorporated into the prohibitive principle of energy justice. What is "due" to people is their enjoyment of those basic goods of human life necessary for a self-respecting existence. To the extent that a coal mine or a power plant or a pipeline interferes with this enjoyment, it must be on the basis that the perceived benefits of this energy infrastructure outweigh the perceived burdens, not measured across all of society, but in terms of their impact on the specific community or communities that are affected.

Two implications follow from these considerations. First, the burdens of a given energy system cannot be imposed on anyone, but must be freely accepted. This is partly because no one is going to accept burdens that are not counterbalanced by perceived benefits; it is also because two of the goods to which people are entitled are the ability to exercise choice in how they live and the ability to participate in the decisions that shape their lives. This freedom of choice is removed if a person is forced to endure a burden without consent. Thus there must be procedural mechanisms that ensure public participation in decisions regarding energy. Second, the burdens and benefits of energy systems will be more or less equitably distributed. An inequitable distribution implies that the imposition of burdens on an individual or a group of people is "undue."

An advantage of the prohibitive principle is that it sidesteps a contested issue that arises in the context of transboundary pollution and global warming – namely, how principles of justice based on need can extend beyond a given community, which in most cases means the nation state. As we have already discussed, the objection to the claims of international justice is that an ethical response to the needs of another *outside the community* should be understood as an act of humanity or compassion, not as a requirement of justice. The prohibitive principle does assume that the entitlements of people to preconditional goods are based on human need, but in the context of global warming and international justice, it only requires industrialized nations to take responsibility for the consequences of their own greenhouse gas emissions, without placing on them any obligation to respond affirmatively to meet the needs of the most vulnerable in developing countries. The distributive principle involved here is that of desert, not need. Those who are most responsible for climate change are likely to suffer least from its effects, while those who have contributed the least to greenhouse gas emissions are most vulnerable to its present and future impacts. In other words, the distribution of the burden of climate change risk is not deserved by those upon whom it falls most heavily.[75]

It is one thing to be asked to respond to the need of another who is a stranger; it is another to be asked to refrain from harming the other undeservedly through our actions. While we might reject the claim that the former instance is a matter of justice, and insist that the only principle involved is benevolence, this position is much harder to justify in the second case. Even if we accept that the claims made by human need do not extend as a matter of justice beyond a given community, no one (presumably) would want to argue that the duty not to harm another likewise does not extend to harming those who are not members of the community.

A variation of this argument can be used to address what some regard as the thorniest issue relating to climate change: the question of our duty to future generations. The basic problem here relates to the notion of reciprocity. Theories of justice based on some version of the social contract require as a minimum condition of justice a relationship of reciprocity to exist between members of the community of justice. In what sense can it be said that there is any reciprocity with people who are not yet born? Strictly speaking, of course, there is no reciprocity in the sense of give and take or exchange between the present generation and future generations. And yet this strict reading of reciprocity would seem to depend on a perspective that regards the present generation as in some way isolated, detached, and complete in itself (which, it might be added, does seem to be the predominant attitude peculiar to modernity). Other ways of regarding the relationship between generations are certainly possible. The statesman Edmund Burke, for instance, expressed a very different sense of the human community: "As the ends of partnership cannot be obtained in many generations, it becomes a partnership not only between those who are living, but between those who are living, those who are dead, and those who are not yet born."[76]

The political philosopher Brian Barry articulated an argument for intergenerational justice that follows along similar lines. "From a temporal perspective," he observes, "no one generation has a better or worse claim than any other to enjoy the earth's resources . . . The minimal claim of equal opportunity is an equal claim on the earth's natural resources."[77] He argues that the present generation cannot claim that it is entitled to a larger share of goods supplied by nature, because most of our technology and capital stock are not the sole creation of the present.[78] As we cannot claim exclusive credit for resources that are inherited, they fall outside any special claims based on the present generation "deserving them." "Since we have received benefits from our predecessors," Barry writes, "some notion of equity requires us to provide benefit for our successors."[79] In other words, we should preserve our resources out of a sense of reciprocity.

Of course, the response to this is that it is impossible to determine what resources future generations will want, as we cannot claim to know what their needs will be. Thus we come back to the argument that human needs are defined within specific cultural and historical contexts. Industrialization and economic growth, after all, do not only result in greenhouse gas emissions and associated climate change; they also generate the new technology and capital stock that will likewise be passed down to our descendants. Indeed, as we shall see in the next chapter, the argument for discounting the costs of future climate change burdens largely depends on the argument that by investing in economic growth we are actually benefiting future generations by making them wealthier. Who is to say that they would want us to do otherwise? This is simply a choice we make based on our preferences, which has little to do with the preferences of those who will come after us. Ultimately, the point at issue is that beyond two or three generations, the future of humanity is an unknown blank for us, and for this reason, it is argued, it is meaningless to speak of having any relationship to it. If we have no relation to distant future generations, then we can have no duties of justice toward them.

There is no denying that the future is unknown. However, as in the case of international justice, so in the case of intergenerational justice, it is going too far to claim that we can know nothing of the human needs of people 100 or 200 years from now. Those goods that Barry calls "vital interests" and Dobson calls "preconditional goods" – in other words, those goods without which any kind of meaningful human existence other than bare subsistence is impossible – are surely universal human goods, and will therefore be needed by future generations.

The economic historian Robert Heilbroner once asked what he called a "terrible question": "Why should I lift a finger to affect events that will have no more meaning for me seventy-five years after my death than those that happened seventy-five years before I was born?"[80] As he remarks, in the end there is no convincing rational answer to this question. "No argument based on reason will lead me to care for posterity or to lift a finger in its behalf."[81] And yet Heilbroner confesses that he is "outraged" by people (he cites some economists) who let the matter end there. The issue, he contends, developing an argument he borrows from the classical economist Adam Smith, is the perspective adopted in responding to this question. So long as we regard the problems brought on by climate change as disinterested rational inquirers, we will be content to argue about discount rates and dispute the likely impacts of varying rises in temperature. But "it is one thing to appraise matters of life and death by the principles of rational self-interest and quite another *to take responsibility for our choice.*"[82] The professor who casually contemplates the demise of humanity ("Suppose that, as a result of using up all the world's resources, human life did come to an end. So what?"[83]) would not personally consign it to oblivion with the same equanimity. It is ultimately a question of individual responsibility.

The prohibitive principle places the weight of justice on the actions of those involved in designing and implementing energy systems; it articulates a duty or obligation, and specifies to whom this obligation belongs. In this respect, as we shall see, it is different from the affirmative principle, which places the emphasis on rights rather than obligations, and leaves in question the matter of *who* is obliged to secure these rights.

The affirmative principle

The Affirmative Principle states that "if any of the basic goods to which every person is justly entitled can only be secured by means of energy services, then in that case there is also a derivative right to the energy service."

As its name implies, this principle "affirms": It establishes a positive right to basic energy services and calls on people to act in providing these services to all of humanity. The affirmative principle follows from a consideration of all four assumptions, with a particular emphasis on the final one, which argues that energy is a material prerequisite for other basic goods. It covers the important issues of energy poverty and fuel poverty, and most importantly, provides an affirmative answer to the question whether it is even possible to speak of having rights to energy services.

At the core of the affirmative principle is the insight articulated powerfully by the late economist E. F. Schumacher, when he pointed out that "[t]here is no substitute for energy. The whole edifice of modern society is built upon it. . . . It is not just another commodity, but the precondition of all commodities, a basic factor equal with air, water, and earth."[84] (Or, as the former U.S. Secretary of Energy Hazel O'Leary once remarked, "Electricity is just another commodity in the same way that oxygen is just another gas."[85]) We have described this as the instrumental domination of energy, and one of its consequences is that energy services are a "material prerequisite" for many of the basic goods of human life, from shelter and education to health services and political representation. Thus, while we recognize that it is, strictly speaking, meaningless to speak of a universal right to something so historically and culturally contingent as modern energy services, we maintain that once the full force of energy's instrumental domination over the means of access to basic goods is properly grasped, then it follows that in certain circumstances (where the instrumental domination of energy is complete) there is a derivative right to energy services. In other words, the success of modern energy systems in pushing out other forms of satisfying human needs – to the point where in some cases it has actually become impossible to satisfy these needs *except through the means of energy services* – has itself created the affirmative right to energy.

Unlike the prohibitive principle of energy justice, which establishes an obligation for the producers and consumers of energy, the affirmative principle establishes a right to energy, and clearly this right most closely concerns those who have the need to assert it. The principle requires us to acknowledge the reality of fuel and energy poverty in the world, which affect billions of people in developed and developing countries; it also requires us to acknowledge that a lack of access to energy services does not merely involve being deprived of such elementary goods as heating, lighting, and transportation, but also deprives people of the ability to develop capabilities that are essential for a flourishing life. Our second assumption about the centrality of capabilities in an adequate conception of human goods is thus necessary to appreciate the full implications of global energy inequality, as it directs our attention, in Martha Nussbaum's phrase, "to what people are actually able to do and to be." In other words, any theory of justice that is concerned with the distribution of such things as civil rights, political freedoms, and equality of opportunity, must necessarily concern itself also with the equitable distribution of energy services. Energy justice is a precondition for the realization of social justice.

It should be clear that our second principle also differs from the first one in that it relies on the distributive principle of need, rather than desert. This is not a problem in the context of meeting domestic needs within the nation state.

The history of modern energy provides many examples of the implicit recognition of this principle, from the New Deal's Rural Electrification Administration to more recent initiatives like Warm Front scheme in England, which addresses the hardships of fuel poverty for qualifying lower-income households.

In these cases, there is no incongruity in appealing to principles of equity and justice. It becomes more of a puzzle in the international context, for all the reasons we have already described. Most people are willing to acknowledge that it is deplorable that over a billion people today have no access to energy services whatsoever, but far fewer people would accept that this is a matter of justice, rather than simply a cause for humanitarian concern: Benevolence, in other words, rather than justice, is the operative principle. Here again we meet the contingent nature of the right to energy services: a hundred years ago it would *not* have been a question of justice that the majority of the world's population had no access to modern forms of energy. And yet today it is. What accounts for the difference?

We shall take up this question in more detail in Chapter 4. Here one word will almost suffice: globalization. Globalization has brought every country (and most of the inhabitants of those countries) within the orbit of the modern industrialized economy. But within modern industrialized societies, as we have already acknowledged, energy is a material prerequisite for meeting many basic human needs. For better or for worse, few people in the world today can escape the domination of energy. It is one thing, of course, to acknowledge this, and another to claim that energy justice requires an international effort to provide an equitable distribution of energy services globally. The affirmative principle, as we have formulated it, leaves in question who has the obligation to provide the necessary energy services. It only stipulates the right to such services. Obviously it is most desirable if the services can be provided domestically; however, seeing that this is not realistic in many cases, there will also be a role for international aid, multinational agencies, and possibly multinational corporations.

We recognize that the affirmative principle does not avoid as easily as the prohibitive principle the debate over the nature of international justice. How the community of need is defined is a real issue in this context. Practically speaking, as our reference to international aid implies, much of the work that is done on this front will take the form of humanitarian initiatives. But this does not negate the fact that the underlying issue involved is what is due to people as a matter of justice. Once we accept that justice requires us to meet certain basic human needs, it follows that energy services (in many circumstances) are also due as a matter of justice.

Conclusion

Global energy systems today both reflect and perpetuate "gross inequalities of access and power."[86] The *prohibitive* and *affirmative* energy justice principles enunciated here offer us a way to move beyond our subjection to the inequitable and destructive aspects of our energy system. The two principles demand that we seriously consider whether it is just that one-quarter of humanity has no access to electricity, and another quarter has less than a tenth of what those of us in industrializing countries had a decade ago. They ask that we decide whether it is just to deplete hundreds of millions of years of energy resources in a few generations, or to reap the benefits of greenhouse gas emissions today at the expense of those not yet born.

As Amartya Sen once observed, any adequate theory of justice will underline the fact that we are all accountable for the decisions that we take. Once we take personal responsibility for the consequences of our individual energy use, and make an effort to imagine the present and future human suffering that we are helping to bring about, we realize that the global energy system cannot, from a moral and ethical perspective, continue on its current pathway. The next five chapters showcase how pervasive and egregious some global energy injustices truly are before the final chapter offers readers a corridor through which they can accept their complicity in this system and begin to change it.

Notes

1. Paul Wilkinson, Kirk Smith, Michael Joffe, and Andrew Haines, "A Global Perspective on Energy: Health Effects and Injustices," *Lancet* 370 (2007): 965–977.
2. Edward A. Page, "Intergenerational Justice of What: Welfare, Resources or Capabilities?" *Environmental Politics* 16, no. 3 (2007): 453–469.
3. Darrel Moellendorf, *Cosmopolitan Justice* (Boulder, CO: Westview Press, 2002), 171.
4. David Miller, *Social Justice* (Oxford: Clarendon Press, 1976), 18.
5. See Ronald Dworkin, "What is Equality? Part 1: Equality of Welfare," *Philosophy and Public Affairs* 10, no. 3 (1981): 185–246; Ronald Dworkin, "What is Equality? Part 2: Equality of Resources," *Philosophy and Public Affairs* 10, no. 4 (1981): 283–345. See also R. Dworkin, *Sovereign Virtue: The Theory and Practice of Equality* (Cambridge, MA: Harvard University Press); and John Rawls, *A Theory of Justice: Revised Edition* (Cambridge, MA: Belknap Press, 1999).
6. Nicholas Georgescu-Roegen, "Energy and Economic Myths," in *Valuing the Earth: Economics, Ecology, Ethics,* ed. Herman E. Daly and Kenneth N. Townsend (Cambridge, MA: MIT Press, 1993), 99.
7. Ibid., 101.
8. Eugene A. Rosa, Gary E. Machlis, and Kenneth M. Keating, "Energy and Society," *Annual Review of Sociology* 14 (1988): 149–172, 149.
9. Ivan Illich, *Energy and Equity* (London: Calder & Boyars Ltd., 1974), 31.
10. Ibid. Illich makes interesting observations on the psychological and cultural implications of modern transport: "The habitual passenger cannot grasp the folly of traffic based overwhelmingly on transport. His inherited perceptions of time and space and of personal pace have been industrially deformed. He has lost the power to conceive of himself outside of the passenger role. Addicted to being carried along, he has lost control over the physical, social and psychic powers that reside in man's feet. The passenger has come to identify territory with the

untouchable land-scape through which he is rushed. He has become impotent to establish his domain" (37).

11. See National Academy of Engineering (NAE), *Greatest Engineering Achievements of the 20th Century* (2003), http://www.greatachievements.org.
12. David Nye, *Consuming Power: A Social History of American Energies* (Cambridge, MA: MIT Press, 1999), 3.
13. Richard F. Hirsh, *Power Loss: The Origins of Deregulation and Restructuring in the American Electric Utility System* (Cambridge, MA: MIT Press, 1999), 38–41.
14. Ibid., 52.
15. See, e.g., John Ash, "New Nuclear Energy, Risk, and Justice: Regulatory Strategies for an Era of Limited Trust," *Politics and Policy* 38, no. 2 (2010): 255–284; Brian Barry, *Justice as Impartiality* (Oxford: Oxford University Press, 1995).
16. Will Hutton, *The World We're In* (London: Abacus, 2002), 87.
17. John Farrington and Conor Farrington, "Rural Accessibility, Social Inclusion and Social Justice: Towards Conceptualization," *Journal of Transport Geography* 13 (2005): 1–12.
18. Darrel Moellendorf, *Cosmopolitan Justice* (Boulder, CO: Westview Press, 2002), 171.
19. See Charles Jones, *Global Justice* (Oxford: Oxford University Press, 1999); and Moellendorf, *Cosmopolitan Justice.*
20. For a compelling take on why justice concerns should apply to non-human species, see Martha C. Nussbaum, *Frontiers of Justice: Disability, Nationality, Species Membership* (Cambridge, MA: Belknap Press, 2006).
21. The Universal Declaration of Human Rights, Article 1, http://www.un.org/en/documents/udhr.
22. Simone Weil, *The Need for Roots: Prelude to a Declaration of Duties Towards Mankind* (London: Routledge Classics, 2002), 6.
23. See Nussbaum, *Frontiers of Justice*, 159.
24. Ibid., 160.
25. Ibid.
26. Ibid., 278.
27. Ibid.
28. Brian Barry, "Sustainability and Intergenerational Justice," in *Fairness and Futurity: Essays on Environmental Sustainability and Social Justice,* ed. Andrew Dobson (Oxford University Press, 1999), 105.
29. Andrew Dobson, *Justice and the Environment: Conceptions of Environmental Sustainability and Theories of Distributive Justice* (Oxford: Oxford University Press, 1998), 75.
30. Edmund Burke, *Reflections on the Revolution in France* (London, 1910), 57.
31. World Commission on Environment and Development, *Our Common Future* (Oxford: Oxford University Press, 1987), 8.
32. Michael Walzer, *Spheres of Justice: A Defense of Pluralism and Equality* (New York: Basic Books, 1983), 84.
33. Ibid., 73–78.
34. Ibid., 28–29.
35. Ibid., 65.
36. Ibid., 83.
37. Ibid., 76.

38. Immanuel Kant, "Toward Perpetual Peace," in *Practical Philosophy: Cambridge Edition of the Works of Immanuel Kant*, trans. M. J. Gregor (Cambridge: Cambridge University Press, 1999), 329–358.
39. Nussbaum, *Frontiers of Justice*, 277.
40. Amartya Sen, *The Idea of Justice* (Cambridge, MA: Harvard University Press, 2009), 402.
41. See David Miller, *Social Justice* (London: Clarendon Press, 1976); see also Kirstin Dow, Roger E. Kasperson, and Maria Bohn, "Exploring the Social Justice Implications of Adaptation and Vulnerability," in *Fairness in Adaptation to Climate Change*, ed. W. Neil Adger, Jouni Paavola, Saleemul Huq, and M.J. Mace (Cambridge, MA: MIT Press, 2006), 83.
42. See Thomas Nagel, "The Problem of Global Justice," *Philosophy and Public Affairs* 33 (2005): 115.
43. Ibid., 130–133.
44. Alan Ryan, *On Politics: A History of Political Thought from Herodotus to the Present* (New York: Penguin, 2012), 146.
45. Sen, *The Idea of Justice*, 26.
46. John Rawls, *A Theory of Justice* (Cambridge, MA: Harvard University Press, 1971), 4, 126.
47. See John Rawls, *The Law of Peoples* (Cambridge, MA: Harvard University Press, 1999).
48. Sen, *The Idea of Justice*, 250.
49. See Sen, *The Idea of Justice*; Nussbaum, *Frontiers of Justice*.
50. Aristotle, *Nicomachean Ethics*, trans. T. Irwin (Indianapolis, IN: Hackett, 1985), 1097b 8–11.
51. John O'Neill, *Ecology, Policy and Politics: Human Well-Being and the Natural World* (London: Routledge, 1993), 88.
52. Sen, *The Idea of Justice*, 322.
53. Ibid., 18.
54. Ibid., 19.
55. Burns H. Weston, "Climate Change and Intergenerational Justice: Foundational Reflections," *Vermont Journal of Environmental Law* 9 (2008): 375–430; see also Burns H. Weston and Tracy Bach, *Climate Change and Intergenerational Justice: Present Law, Future Law* (Vermont Law School, 2008).
56. Jouni Paavola, W. Neil Adger, and Saleemul Huq, "Multifaceted Justice in Adaptation to Climate Change," in *Fairness in Adaptation to Climate Change*, ed. W. Neil Adger, Jouni Paavola, Saleemul Huq, and M. J. Mace (Cambridge, MA: MIT Press, 2006), 263–277.
57. Nussbaum, *Frontiers of Justice*, 77.
58. Donald K. Anton and Dinah L. Shelton, *Environmental Protection and Human Rights* (New York: Cambridge University Press, 2011), 431.
59. Robert Goodland, "Free, Prior and Informed Consent and the World Bank Group," *Sustainable Development Law and Policy* 4, no. 2 (2004): 66–74; UN Permanent Forum on Indigenous Issues (UNPFII), Report of the International Workshop on Methodologies Regarding Free Prior and Informed Consent and Indigenous Peoples (Document E/C.19/2005/3), submitted to the Fourth Session of UNPFII, May 16–17, 2005.
60. Goodland, "Free, Prior and Informed Consent."

61. Matt Finer, Clinton N. Jenkins, Stuart L. Pimm, Brian Keane, and Carl Ross, "Oil and Gas Projects in the Western Amazon: Threats to Wilderness, Biodiversity, and Indigenous Peoples," *PLoS One* 3, no. 8 (2008): 1–9.

62. Commission on Human Rights, Sub-Commission on the Promotion and Protection of Human Rights, Working Group on Indigenous Populations, twenty-second session, July 19–13, 2004, 5.

63. Marcus Colchester and Maurizio Farhan Ferrari, *Making FPIC – Free, Prior and Informed Consent – Work: Challenges and Prospects for Indigenous People,* Forest Peoples Project, June 4, 2007; United Nations, *Free Prior Informed Consent and Beyond: The Experience of IFAD* (Geneva: PFII/2005/WS.2/10, 2005); Debra J. Salazar and Donald K. Alper, "Justice and Environmentalisms in the British Columbia and U.S. Pacific Northwest Environmental Movements," *Society and Natural Resources* 24, no. 8 (2011): 767–784.

64. See, e.g., Hirsh, *Power Loss*; Joseph P. Tomain and Richard D. Cudahy, *Energy Law,* 2nd ed. (St. Paul: Thomson Reuters, 2011), 52–106.

65. Eugene A. Rosa, Gary E. Machlis, and Kenneth M. Keating, "Energy and Society," *Annual Review of Sociology* 14 (1988): 149–172.

66. See, e.g., Amory Lovins, *Soft Energy Paths: Towards a Durable Peace* (San Francisco: Friends of the Earth International, 1977).

67. See, e.g., Tomain and Cudahy, *Energy Law*, 104.

68. B. K. Sovacool and M. A. Brown, "Conclusions – Replacing Myths with Maxims: Rethinking the Relationship Between Energy and American Society," in *Energy and American Society: Thirteen Myths,* ed. B. K. Sovacool and M. A. Brown (New York: Springer, 2007), 359.

69. Ibid., 361–362.

70. See Lewis Mumford, "Authoritarian and Democratic Technics," *Technology and Culture* 5 (1964): 2–12; Carroll Pursell, "The Rise and Fall of the Appropriate Technology Movement in the United States, 1965–1985," *Technology and Culture* 34, no. 3 (1993): 629–637.

71. E. F. Schumacher, *Small Is Beautiful: Economics as if People Mattered* (London: Blond & Briggs, 1973), 163.

72. Nussbaum, *Frontiers of Justice*, 289.

73. Ibid.

74. Andrew Szasz, *Ecopopulism: Toxic Waste and the Movement for Environmental Justice* (Minneapolis: University of Minnesota Press, 1994), 151.

75. This argument has been adapted from the discussion in Kirstin Dow, Roger E. Kasperson, and Maria Bohn, "Exploring the Social Justice Implications of Adaptation and Vulnerability," in *Fairness in Adaptation to Climate Change,* ed. W. Neil Adger, Jouni Paavola, Saleemul Huq, and M. J. Mace (Cambridge, MA: MIT Press, 2006), 82.

76. Edmund Burke, *Reflections on the Revolution in France* (London: Penguin, 1790/1982), 194–195.

77. Brian Barry, "Justice as Reciprocity," in *Democracy, Power and Justice, Essays in Political Theory* (Oxford: Clarendon Press, 1989), 463–494.

78. Brian Barry, "Circumstances of Justice and Future Generations," in *Obligations to Future Generations,* ed. Richard Sikora and B. M. Barry (Philadelphia, PA: Temple University Press, 1978).

79. Barry, "Justice as Reciprocity," 403.

80. Robert Heilbroner, *An Inquiry into the Human Prospect: Looked at Again for the 1990s* (New York: W.W. Norton & Company, 1991), 184.
81. Ibid.
82. Ibid., 188 (emphasis added).
83. Ibid., 184.
84. E. F. Schumacher, *Schumacher on Energy: Speeches and Writings of E.F. Schumacher,* ed. G. Kirk (London: Cape, 1977), 1–2.
85. Quoted in Ralph Cavanagh, "Restructuring for Sustainability: Toward New Electric Service Industries," *Electricity Journal* (July 1996): 71.
86. Nussbaum, *Frontiers of Justice*, 312.

3 The temporal dimension

Externalities and climate change

Climate change is the most regressive tax in the world: the poorest pay for the actions of the rich.
— Kirk R. Smith, quoted in "Climate Change and the Poor: Adapt or Die," *The Economist* (September 11, 2008), 51

Introduction

In his statement, U.C. Berkeley Professor Kirk R. Smith provocatively argues that climate change has pressing moral and ethical concerns in addition to economic and environmental ones. As he suggests, climate change essentially involves marginalized and geographically distant communities such as those in Bangladesh or Benin "paying" for the high levels of energy consumption of wealthy consumers in Berlin and Beverly Hills. Moreover, the severity and scope of many of these impacts will shift in time: Some are immediate, whereas others are temporally distant.

Professor Smith's statement brings to mind a central question in discussions about energy and climate policy: what is the present value of future damages from climate change? The controversy surrounding the 2006 *Stern Review on the Economics of Climate Change* mostly centered on the low discount rate adopted by the report – it was, in other words, an argument about time: How much weight to give to the welfare of future generations compared to the welfare of present generations.[1] The "social rate of time discount" is a means of comparing the present costs of abating carbon dioxide emissions with the future benefits of reducing the damages of climate change, understood comprehensively to include market, nonmarket, and ecological impacts. Much of the debate concerning climate change policy revolves around the question of what discount rate should be used to calculate the value of future benefits and costs discounted to the present. What is sometimes lost in these abstract considerations is any thought for the real people, both present and future, for whom these costs will be more than simple numbers on a page.

The justification for using discounting as a tool for thinking about climate change policy is based on two assumptions. First, that "most" climate damages to be avoided "lie far in the future."[2] Second, that the kind of compound economic growth we have seen since the eighteenth century, and especially over the past half century, will continue indefinitely into the future. Hence, as one article put it, "[f]uture costs are discounted because the future world will be richer and better able to afford them. Future benefits are discounted because they will be a diminishing fraction of future wealth."[3] According to its critics, the Stern Report overestimated future economic costs of climate change and was alarmist in calling for prompt action because it relied on a low discount rate that did not reflect the real interest and savings rates of the market.

The argument advanced by these critics – greatly simplified – goes something like this: economic growth requires the use of fossil fuels to provide the energy needed for industrialization and trade (transportation). If countries make appropriate investments in the economy today, rather than slowing economic growth by reducing the consumption of fossil fuels, future generations will be the beneficiaries of a much wealthier world and, as a result, will be much better positioned than we are to address the adverse impacts of climate change. As Freeman Dyson commented while speaking of China and India, the "decision to become rich" is the most sensible and ethically sound option for countries – one that, economically, justifies the use of a high discount rate.

The flaw in this argument is that it treats the economy as though it were independent of the natural resources upon which it is based, and it presumes that there are no limits constraining the substitutability of manufactured capital for natural capital. And yet for more than 40 years, since at least the 1972 publication of *Limits to Growth* by the Club of Rome, it has become increasingly evident that human economic activities are exceeding the earth's carrying capacity – its ability to regenerate resources and absorb wastes.[4] Resource scarcity (a phrase that includes everything from freshwater, metals, phosphorus, and fish to topsoil, fossil fuels, forests, and rare earth elements) is now becoming a limiting factor for global economic activity. Spiking prices for commodities and fossil fuels in recent years provided a foretaste of what is to come. With global demand increasing at exponential rates, and supplies of many essential resources diminishing, there are good reasons to call into serious doubt the assumption of continuing economic growth.[5] When we also factor in current, probable, and possible impacts of climate change on freshwater resources, arable land, and fish stocks, the conviction that future generations will be wealthier than the present begins to appear fantastical. But once the promise of ever-increasing prosperity (and corresponding resourcefulness) is shaken, the current practice of deferring meaningful action on reducing CO_2 emissions becomes much harder to defend, at least if we have any concern for future generations. From the perspective of intergenerational justice, discounting the future impacts of climate change is little more than a clever sleight of hand that serves to conceal an appalling indifference to the assured suffering of millions of people yet to be born.

As already noted, the other assumption that justifies continued reliance on fossil fuels is that the *serious* adverse consequences of climate change are still a long way off – at least another hundred years – so we still have time to come up with a technological solution. There is an element of truth to this assumption, in that the *catastrophic* scenarios that might be precipitated by climate change – like the collapse of the Greenland ice sheet, which would raise sea levels worldwide by 23 feet, and could halt the Gulf Stream, switch off the Asian monsoon, and warm the South Pacific Ocean, possibly destabilizing another huge ice sheet in West Antarctica – will most likely not be seen for decades or centuries, if at all. And yet (leaving aside the possibility that various tipping points are passed that dramatically increase the speed with which the Earth's climate changes), this assumption completely ignores the fact that climate change is *already* causing enormous destruction of life and property throughout the world. Climate change has a present as well as a future tense. Moreover, the impacts of climate change disproportionately affect the poorest and most vulnerable communities in the developing world – in other words, those who gain least from the energy systems responsible for climate change, the benefits of which are supposed to justify continued reliance upon them.

Unfortunately, modern conventional energy systems have many negative externalities associated with them, not only climate change. Whether they are based on fossil fuels or nuclear power (or even, though to a lesser extent, renewable energy like hydro and wind), every phase of energy extraction, production, and use has associated waste products. These externalities have both immediate impacts on the local environment and human populations (like those associated with emissions of sulfur dioxide), and more delayed impacts that accumulate over time and can last for centuries (as is the case with CO_2 emissions), or even, in the case of radioactive waste, for millennia. Indeed, the late physicist Alvin M. Weinberg once joked that it would be fair to call some energy systems "immortal" since they distribute benefits and costs long after they have been amortized.[6] Some dams, for example, have been producing energy for hundreds of years, and the byproducts of nuclear reactors will last for tens of thousands of years into the future. Furthermore, as Figure 3.1 illustrates, many of the things we build today, including railways, buildings, and power plants, will last upward of 75 to 100 years, meaning they will continue to produce the externalities described next for longer than most of our lifetimes.[7] In other words, our global energy systems rely on "thermodynamic time bombs" that extend their externalities far into the future.[8]

The seemingly "far off" nature of some energy-related externalities, therefore, is a poor justification for complacency with the status quo. The *present* costs of modern energy systems cannot be discounted – at any rate – and as we shall see in this chapter, the burden of these costs on human lives today is enormous. Thus, even if we disregard future damages, there are adequate reasons to fundamentally question the dominant model of energy policy. Unfortunately, what binds the present and future costs of climate change together is the fact that they are and will be borne largely by those who receive no benefit from fossil fuel-created wealth and whose interests are not represented in the boardrooms and legislative chambers of the world: the poor and the not yet born.

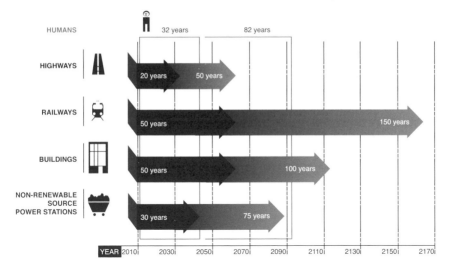

Figure 3.1 Life spans for various energy technologies

Source: United Nations Economic and Social Commission for Asia and the Pacific.

Awareness of this should lend gravity to our deliberations about how to design new energy systems that do not violate fundamental principles of justice.

Externalities and energy burdens

The pervasive environmental problems associated with industrial capitalist economies result from what economists call "market failures." Markets "externalize" environmental and social costs (e.g., hazardous working conditions) and fail to provide or adequately value public goods (e.g., clean air). As geographer David Humphreys observes:

> Most global environmental problems, such as acid rain, global warming and deforestation, are caused by what Julian Saurin terms the "normal and mundane practices" of modern capitalism, such as industrial production and natural resource consumption. Environmental problems are not, therefore, exceptional or accidental; they are the cumulative result of routine social actions.[9]

Thus many externalities result from extracting, producing, and using energy fuels, yet these costs are not reflected in the rates and prices for energy services. What is an "externality"? This is a term used by economists to describe an economic "side effect" – that is, an unintended cost or benefit resulting from an economic activity that affects people other than those engaged in the economic activity.[10] An externality can be either "positive" or "negative." For instance, when apple growers happen to live adjacent to honey producers, the bees pollinate the apple trees, just as the apple blossoms increase the production of honey at the bee farm. Either result could be regarded as a positive externality. A negative externality, on the other hand, might be suffered by a woman who supports herself by hand laundry living adjacent to a new factory. Now, when she hangs the clothes out to dry, they are blackened by the factory smoke. The uncompensated loss of income is the negative externality.[11]

Discussions of negative externalities date all the way back to antiquity. When ancient Rome burned wood during the cold winter months, Emperor Nero's tutor Seneca complained of the bad effect the smoke had on his health and of smoke damage to temples. King Edward I issued a proclamation in 1306 that the use of coal was punishable by death, and at least one man was executed under this law. A seventeenth-century court case in England involved a pig farm that polluted neighbors' lands with waste and noxious fumes and as a result lowered property values.[12] Many nineteenth-century lawsuits throughout Europe and the United States involved railroads: Sparks from locomotive trains caused brush fires on farms; exposure to the coal smoke from locomotives increased mortality rates for tuberculosis and pneumonia; and infrastructure around railways had to be regularly cleaned, painted, and replaced.[13] One early nineteenth-century case involved the death of 1,000 cattle, 800 sheep, and 20 horses during the first year of operation of a smelter.[14]

Rational firms will usually overproduce negative externalities (since somebody else pays for them) but under-produce positive externalities (since they are prone to free riders).[15] While nineteenth-century classical economists were very cognizant of the virtue of the market as an efficient mechanism for the allocation of scarce resources, they understood that it could only operate satisfactorily within a framework of legal, political, and moral restrictions. Left to their own devices, firms would inevitably produce externalities in the interest of profit and growth.[16] When the principles of neoclassical economics were being formulated by Alfred Marshall in the 1890s and Arthur Cecil Pigou in the early 1900s, one of their central arguments was that externalities *had* to be internalized (or taxed, to use Pigou's language). Otherwise, firms would always exploit the system to shift as many private costs as they could to the public.[17]

Approximately half a century later, Garrett Hardin developed the term "tragedy of the commons" to describe how people (and firms) externalize the costs associated with their use of "common" or more accurately "open access" resources, under the rational expectation that if they do not, others will.[18] Hardin references leased agricultural grazing lands and the National Parks as examples of the commons. In each case, the common resource has a tendency to be exploited because the benefits of exploitation accrue to a small number of individuals, whereas the costs are distributed to everyone. As Hardin notes, "we are locked into a system of fouling our own nest, so long as we behave only as independent, rational, free-enterprisers."[19]

Two common resources exploited by conventional energy systems are clean air and clean water. Coal-fired power plants use the air as an open sink for carbon dioxide, sulfur dioxide, nitrous oxides, mercury, particulate matter, and ozone – to name only some of the pollutants – shifting to society and the natural environment the costs of acid rain, climate change, mercury poisoning, asthma, lung cancer, and so on. Coal mining also externalizes environmental costs, polluting streams with mine spoils and degrading water quality for local communities. Offshore oil production results in millions of gallons of contaminated water containing lead, zinc, mercury, and benzene being dumped daily into the ocean, with inevitable impacts on local biodiversity and human health. Uranium mining exposes miners to excessively high levels of radon, leading to many premature deaths by lung cancer.

To be fair, the energy services provided by conventional energy systems generate unprecedented opportunities for those who have access to them; yet the externalities that result from these same systems limit opportunities for many others, frequently to the extent of making it difficult to live a meaningful and healthy human life, and sometimes to the extent of making it impossible to live. The prohibitive principle of energy justice states that energy systems should be designed so as not to interfere unduly with the ability of any human being to acquire the basic goods to which he or she is justly entitled. Another way of stating the prohibitive principle would be to say that energy systems should be designed so as to minimize externalities. Conventional energy systems, unfortunately, appear to be designed to do the opposite, and maximize externalities.

The list of externalities associated with conventional energy systems goes on and on. For instance, the extravagant consumption and contamination of freshwater by fossil fuel-based energy systems is an increasingly pressing and urgent concern in a world with diminishing resources of freshwater, yet we only deal with the subject indirectly in what follows, simply for lack of space. Though far from an exhaustive list, in this chapter we elaborate on four specific types of energy-related externalities to provide a sense of their scale and scope: climate change, the externalities associated with extracting fossil fuels and uranium, the legacy of nuclear waste, and air pollution and respiratory diseases.

Climate change

Climate change, as the economist Richard Tol surmised, might well be "the mother of all externalities."[20] Not only are the costs of energy production shifted to the entire globe, but they will also extend far into the future. Climate change will affect nearly every living creature, now and for millennia to come. These costs, need it be said, are not reflected in the price of fuel.

According to the Intergovernmental Panel on Climate Change (IPCC), human sources emitted 49 billion metric tons of carbon dioxide equivalent (CO_2e) into the atmosphere in 2004 – the last time a global inventory was taken.[21] Global greenhouse gas (GHG) emissions grew 70 percent from 1970 to 2004 and, if trends continue, could increase 130 percent by

2040. According to the most recent data for the United States, total emissions amounted to 6.8 billion tons of CO_2e, an increase of 3.2 percent over 2009 levels and an increase of 10.5 percent from 1990 levels. The U.S. Environmental Protection Agency (EPA) attributes such increases to economic growth, rising demand for electricity, and warmer summer weather.[22]

Despite all of the political and media attention surrounding climate change and global warming, carbon-intensive fuels continue to dominate worldwide electricity generation. Fossil fuel power plants in the United States, for instance, emit more than 10 times the amount of CO_2 than the next-largest emitter – iron and steel production.[23] China currently constructs one coal plant each week, and other developing countries – led by India – are gearing up to reach about one-third of that level of coal plant production by 2030. David Hawkins and his colleagues from the Natural Resources Defense Council estimate that new coal-fired power plants, over their 60-year life spans, could collectively introduce into the atmosphere approximately as much CO_2 as was released by all of the coal burned since the dawn of the Industrial Revolution.[24]

Aside from power plants, the other main source of global CO_2 emissions is oil-fueled transportation. With the rise of an affluent middle class in China and India, the demand for automobiles is increasing. Global trade is almost completely dependent on petroleum products. In the case of air transport, it is commonly agreed that airplanes have more or less made all the efficiency improvements that the laws of physics will allow.[25] Yet air travel remains a major source of greenhouse gas emissions.

In recent years, high fuel prices and innovative technology have opened up large reserves of "shale gas" – natural gas extracted from gas shales, or porous rocks that hold gas in pockets. Natural gas is often touted as a "cleaner" fuel than coal or oil, and an ideal "transitional" fuel to an energy system based on renewable sources of energy, and yet the "shale revolution" is by no means an unmixed blessing. Shale gas is captured by hydraulic fracturing, or "fracking," a process that shatters rocks by injecting water to release the gas. *Nature* cautions that 0.6 to 3.2 percent of the methane can escape directly during the production process; as methane is 70 times more powerful at heating the atmosphere than carbon dioxide over a 20-year time frame, this is especially worrying (although it should also be noted that methane remains in the atmosphere for a much shorter period than CO_2, persisting for only 12 years approximately).[26]

Studies on "fugitive" methane from shale gas production, however, have showed conflicting results. One study by MIT researchers concluded that the amount of methane emissions has been exaggerated.[27] Other studies, though, have noted that 3.6 to 7.9 percent of methane from shale gas production escapes to the atmosphere in venting and leaks, which if accurate would make methane emissions from shale gas production as much as twice as great as methane emissions from conventional natural gas production and even worse than coal generation – see Figure 3.2.[28] Processing, where hydrocarbons and impurities such as sulfur are removed, is energy and carbon

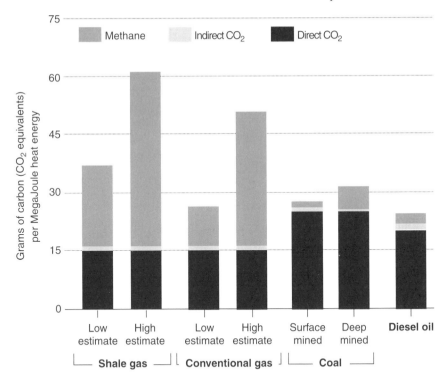

Figure 3.2 Lifecycle greenhouse gas emissions for shale gas, conventional gas, and coal, 2011

Source: Robert W. Howarth and Anthony Ingraffea.

intensive as well, and shale gas needs to be more extensively processed to make it ready for existing pipelines. Furthermore, the international trade in natural gas requires it to be shipped as Liquefied Natural Gas (LNG), and the process of liquefaction, whereby the gas is cooled to a temperature of $-162\,^{\circ}C$, is highly energy intensive.

Some energy analysts regard greater reliance on nuclear energy as the most effective way to reduce CO_2 emissions. And yet, despite recent claims by the Nuclear Energy Institute that nuclear power is "the Clean Air Energy," nuclear power is responsible for more greenhouse gas emissions than most people realize. Reprocessing and enriching uranium requires a substantial amount of electricity, often generated from coal-fired power plants, and uranium milling, mining, leeching, plant construction, and decommissioning are all greenhouse gas intensive. When one takes into account the carbon-equivalent emissions associated with the entire nuclear life cycle, nuclear plants do contribute significantly to climate change. For instance, one assessment of 103 lifecycle studies of carbon dioxide equivalent emissions for nuclear power plants found that the average CO_2 emissions over the typical lifetime of a plant are about 66 grams for every kWh, or the equivalent of some 190 million metric tons of CO_2 in 2010.[29] The Oxford Research Group projects that because of the inevitable eventual shift to lower quality uranium ore, if the percentage of world nuclear capacity remains what it is today, by 2050 nuclear power would generate as much carbon dioxide per kWh as comparable natural gas-fired power stations.[30]

In aggregate, the damages from carbon dioxide emissions – something not currently factored into most energy markets – could be staggeringly large, exceeding the expense of previous global threats such as AIDS, the Great Depression, and global terrorism. Researchers from the Pew Center on Global Climate Change warn that "waiting until the future" to address global climate change might bankrupt the U.S. economy.[31] The Stern Report projected that the overall costs and risks of climate change will be equivalent to losing at least 5 percent of the world's GDP, or $3.2 trillion, every year, now and forever, and that these damages could exceed 20 percent of GDP ($13 trillion) if more severe scenarios unfold.[32] (As we already commented, the Stern Report generated much controversy, some saying that it was too conservative, and others that it exaggerated the economic impacts of climate change. Stern himself recently observed that he "got it wrong on climate change – it's far, far worse."[33]) For example, global economic damages from extreme weather events, only one type of externality related to climate change, have doubled every 10 years and reached about $1 trillion in total damages over the past 15 years. Annual weather-related disasters have increased by a factor of 4 from 40 years ago, and insurance payouts have increased by a factor of 11 over the same period, rising by $10 billion per *year* for most of the past decade.[34]

The World Health Organization (WHO) believes that by 2000 climate change had already killed 150,000 people and subjected a further 5.5 million people to years of lost life due to debilitating diseases. Most of these people have been in the developing world. In the Maldives, changing precipitation patterns and flooding have contributed to water- and vector-borne diseases and the increased morbidity and mortality that accompany them. Waterborne diseases such as shigellosis and diarrheal diseases have also become more pronounced in children under the age of five, spread by an increase in flooding. Indirectly, climate change has contributed to malnutrition and limited the accessibility and quality of health care, with storms and floods making it more difficult to distribute food or transport patients to doctors.[35] During the monsoon season in 2004 in Bangladesh, for example, flooding placed 60 percent of the country under a solid pool of water that was mixed with industrial and household waste. More than 20 million people were affected and many suffered shortages of water, skin infections, and communicable diseases.[36]

More worryingly, the WHO projects at least a *doubling* in deaths and burden in terms of life years lost by 2030, due to heat-related illnesses, illnesses from floods, droughts, and fires, and changing patterns in vector-borne diseases and increasing loss of biodiversity.[37] In China, climate change will in all likelihood alter disease vectors and create conditions for pandemics, as increases in temperature and decreases in availability of water expand the range and frequency of malaria, dengue fever, and encephalitis.[38] In low-lying river deltas, flooding and cyclones will directly affect health and nutrition by causing physical damage and disruptions in the supply of food and basic services, and will have indirect consequences by spreading water-borne diseases and creating prolonged periods of malnutrition.

The United States Department of Defense has simulated the probable effects of climate change and has begun preparing for a future world where droughts and periods of extreme heat increase in the southwestern United States and Mexico; where hurricane intensity increases on the coast of the United States and the Caribbean basin; where ice storms become more difficult to deal with in New England and Eastern Canada; where large mudslides and flooding occur in Central America; where massive wildfires cause deforestation, flooding, and siltification not only in California, Washington, and British Columbia, but also Argentina and Brazil; and where increasingly violent typhoons and cyclones affect the Philippines, India, Bangladesh, Vietnam, and China, severely damaging coastal cities.[39]

Climate change is increasing the frequency and severity of natural and humanitarian disasters. In Columbia, for example, changing precipitation patterns, more extreme weather events, hurricanes and flooding, stronger cycles of El Niño and La Niña, and increases in sea levels threaten both coasts and are expected to produce notable hardships for major population centers.[40] In the Maldives, about half (44 percent) of all human settlements and 70 percent of all critical infrastructure are within 100 meters of the sea. These settlements are already vulnerable to rising sea levels, storms, and floods. Severe weather events from 2000 to 2006 flooded 90 inhabited islands at least once and 37 islands regularly. Sea swells in 2007 inundated 68 islands in 16 atolls, destroyed 500 homes, and necessitated the evacuation of 1,600 people.[41]

In Bolivia and parts of Latin America, flooding from storms is expected to contribute to landslides that result in higher rates of morbidity and mortality.[42] Mountainous areas around the world from the Alps in Europe to the Himalayas in Asia also face the aggravated risk of glacial lake outburst floods – when glaciers melt faster than expected and produce massive, spontaneous releases of water capable of killing thousands of people and destroying entire cities. The United Nations Environment Program and the International Center for Integrated Mountain Development have identified no less than 24 high-risk glacial lakes near Bhutan and Nepal.[43] Melting glaciers could flood river valleys in Kashmir and Nepal, endangering the lives of 182 million people with disease epidemics and starvation.[44] In Africa, rising sea levels could destroy as much as 30 percent of the continent's entire coastal infrastructure.[45]

The impacts from climate change just described are only a somewhat arbitrary selection, and even then, we have focused on those damages that are either already occurring or almost certain to occur. Runaway climate change, which could result from passing beyond various tipping points identified by climate scientists, like a massive release of methane hydrates from the arctic tundra and the ocean, would change the picture completely. Whether we focus on the present or the future, climate change presents us with the prospect of human suffering quite literally unprecedented in all of history, with entire nations sinking beneath the waves, millions of uprooted people, and millions more with inadequate access to water.

While a changing atmosphere will eventually affect us all, it is also true that for the time being climate change impacts are having a disproportionate impact on people within developing countries that are most vulnerable to them. Those of us who enjoy the benefits of modern energy systems, such as air conditioning, motorized mobility, and mobile phones and laptops, are for that reason able to insulate ourselves to some degree from the externalities of our activities. The justice implications of this state of affairs are too obvious to be spelled out.

Extracting fossil fuels and uranium

How we extract fossil fuels and uranium – how we dig for coal, drill for oil and gas, and leech uranium out of the ground – involves an almost perverse collection of externalities, with huge costs in terms of degraded ecosystems and diminished human health. The damages that result from extraction processes have both an *immediate* impact, often on the people involved in the extracting process and on local communities and ecosystems, and a *deferred* impact, which in the case of uranium mining extends deep into the future.

Many stages of the oil and gas fuel chain – exploration, onshore and offshore drilling, refining – pose serious and unavoidable risks extending beyond those using or profiting from the oil and gas. Exploration necessitates heavy equipment and can be quite invasive, as it involves "discovering" oil and gas deposits found in sedimentary rock through various seismic techniques such as controlled underground explosions, special air guns, and exploratory drilling.[46] The construction of access roads, drilling platforms, and their associated infrastructure frequently leads to environmental impacts beyond the immediate effects of land clearing, by opening up remote regions to loggers and wildlife poachers. For every one kilometer of new oil and gas roads built through forested areas, about 1,000 to 6,000 acres of land are deforested; in Ecuador, an estimated 2.5 million acres of tropical forest were colonized due to the construction of just 500 kilometers of roads for oil production.[47]

The extraction and production of oil and gas is even more hazardous.[48] Drilling for oil and gas involves bringing large quantities of rock fragments, called "cuttings," to the surface, and these cuttings are coated with drilling fluids, called "drilling muds," which operators use to lubricate drill bits and stabilize pressure within oil and gas wells. The quantity of toxic cuttings and mud released for each facility is gargantuan, ranging between 60,000 to 300,000 gallons per day. In addition to cuttings and drilling muds, vast quantities of water contaminated with suspended and dissolved solids are also brought to the surface, creating what geologists refer to as "produced water."[49] The average offshore oil and gas platform in the United States releases about

400,000 gallons of produced water back into the ocean or sea every day. Produced water contains lead, zinc, mercury, benzene, and toluene, making it highly toxic; operators are often required to treat it with chemicals, increasing its salinity and making it fatal to many types of plants, before releasing it into the environment. The ratio of waste to extracted oil is staggering: every one gallon of oil brought to the surface yields eight gallons of contaminated water, cuttings, and drilling muds.[50]

The U.S. Geological Survey (USGS) estimated that there are more than 2 million oil and natural gas wells in the continental United States. But the most intense areas of oil and gas production are off the shores of the Gulf of Mexico and along the northern coast of Alaska. Offshore oil and natural gas exploration and production in the Gulf of Mexico exposes aquatic and marine wildlife to chronic, low-level releases of many toxic chemicals through the discharge and seafloor accumulation of drilling muds and cuttings, as well as the continual release of hydrocarbons and polluted water around production platforms.[51] One study found that mercury levels in the mud and sediment beneath the oil platforms in the Gulf of Mexico were 12 times greater than acceptable levels under federal EPA standards.[52] Exposure to these chronic environmental perturbations continues to threaten marine biodiversity and human health over wide areas of the Gulf, to say nothing of when accidents such as *Deepwater Horizon* occur. The Natural Resources Defense Council also noted that the onshore infrastructure required to sustain oil and natural gas processing in the United States has destroyed more coastal wetlands and salt marshes than can be found in the total coastal area stretching from New Jersey through Maine. Similarly, the U.S. Minerals Management Service has documented 70 oil and natural gas spills between 1980 and 1999 from just the production of oil and natural gas. During this period, oil and natural gas spills accounted for more than 3 million gallons of fuel released into the Gulf of Mexico.[53]

The extraction of coal imposes similar externalities on communities and ecosystems near mining sites. Coal mining, for instance, can remove entire mountaintops in a process that clears forests and topsoil before it uses explosives to break up rocks. Mine spoils are pushed into adjacent streams and valleys, which can cause acid drainage into river systems. Of the more than 1 billion tons of coal mined in the United States annually, roughly 30 percent comes from mountaintop removal (see Figure 3.3).[54] The overall process has destroyed ecosystems, blighted landscapes, and diminished the water quality of rural communities. In the Appalachian region of the United States, mountaintop mining with valley fill operations (MTVF) has so far converted 1.1 million hectares of forest into surface mines and buried more than 2,000 kilometers of freshwater streams and rivers. One recent study observed "there is, to date, no evidence to suggest that the extensive chemical and hydrologic alterations of streams by MTVF can be offset or reversed by currently required reclamation and mitigation practices."[55] Another survey of 78 MTVF streams in the same region found that 73 of them had water pollution levels far past the threshold for toxic bioaccumulation.[56] As one of the authors of this study

Figure 3.3 Mountaintop removal coal mining near Kayford Mountain, West Virginia, 2007
Source: U.S. Environmental Protection Agency.

stated, "[T]he scientific evidence of the severe environmental and human impacts from mountaintop mining is strong and irrefutable. Its impacts are pervasive and long lasting and there is no evidence that any mitigation practices successfully reverse the damage it causes."[57] Federal agencies in the United States had to spend $2.8 billion of public money cleaning up abandoned mine sites from 1997 to 2008 – the benefits of harvesting those sites has long since accrued to a select number of firms, while its costs are still borne by society.[58]

Such damages from coal mines are obviously not limited to the United States. One global assessment of the coal mining industry noted that common "direct" impacts include:

> fugitive dust from coal handling plants and fly ash storage areas; pollution of local water streams, rivers, and groundwater from effluent discharges and percolation of hazardous materials from the stored fly ash; degradation of land used for storing fly ash; and noise pollution during operation [in addition to] impacts on the health, safety, and well-being of coal miners; accidents and fatalities resulting from coal transportation; significant disruption to human life, especially in the absence of well-functioning resettlement policies; and impacts on environment, such as degradation and destruction of land, water, forests, habitats, and ecosystems.[59]

Another recent survey of global mining practices concluded that "a serious history of mining accidents" exists largely due to "widespread neglect of environmental safety and human security issues" and "sub-standard management activities"; it also noted a rising amount of transboundary pollution associated with mining and mineral prospecting and that more mines are being opened in states with weak regulatory and governance regimes.[60] A similar World Bank-backed study of mining practices noted that they "often have substantial environmental impacts," negatively impact food security and the collection of clean water, and add greater healthcare and time burdens to women.[61]

In Australia, the world's largest exporter of black coal and the fourth largest producer overall, an assessment of three coal mines – Hunter Valley in New South Wales, the Bowen Basin in Queensland, and the Gunnedah Basin in New South Wales – documented that mining activities have devastated local communities and wrought havoc in ecosystems. It noted that the impacts of multiple mining operations have compounded social and environmental impacts in ways that made safeguards for individual mines ineffective. It also noted that such cumulative impacts extended "well beyond the geographic location of an operation and may contribute to systems already impacted by other operations, industries, and activities.[62] In China, the human and environmental costs of coal mining are even more obscene. Approximately 6,000 miners perish in underground coal mines every year, and the Chinese government has calculated that the costs of environmental damage from coal mining – including wasted resources, environmental pollution, ecological destruction, and surface subsidence – exceed $5 billion per year.[63]

The process of uranium mining is also remarkably wasteful and deleterious to the environment, regardless of the technique employed. As the American Society for Bioethics and Humanities has noted:

> When nuclear power is touted as a "clean" form of energy, proponents are generally referring to the fact that it produces fewer greenhouse gas emissions than fossil fuel. Belied by this rosy description are the enormous and documented ecological and human risks and harms associated with extraction, processing, enrichment, waste storage, and nuclear accidents, as well as with uranium's potential use in weapons production. If we go forward with nuclear power, any morally tenable nuclear energy policy must incorporate enforceable human rights protections into the uranium extraction process.[64]

For example, to produce the 25 tons of uranium needed to keep a typical reactor fissioning atoms for one year, 500,000 tons of waste rock and 100,000 tons of mill tailings – toxic for hundreds of thousands of years – will be created, along with an extra 144 tons of solid waste and 1,343 cubic meters of liquid waste.[65] Underground mining presents a "significant danger," as the radionuclides uranium-235, radium-226, radon, and strontium-21 accumulate in the soil and silts around uranium mines, often inhaled by miners in the form of radioactive dust.[66] Open-pit mining is prone to sudden emissions of radioactive gases and the degradation of land, as kilometer-wide craters are formed around uranium deposits, which interfere with the flow of groundwater as far as 10 km away.[67] Uranium miners perform in-situ leaching, a third technique, by pumping liquids into the areas surrounding uranium deposits. These liquids include acid or alkaline solutions to weaken the calcium or sandstone surrounding uranium ore. Operators then pump the uranium up into recovery wells at the surface, where it is collected. In-situ leaching is more cost effective than underground mining because it avoids the significant expense of excavating underground sites and often takes less time to implement. All three types of uranium mines have been shown to release harmful rates of gamma radiation. At five separate mines in Australia – Nabarlek, Rum Jungle, Hunter's Hill, Rockhole, and Moline – gamma radiation levels exceeded safety standards in some cases by 50 percent, leading to "chronic" exposure to miners and workers.[68]

Such mining produces a variety of negative environmental impacts. The most direct are occupational hazards. For instance, uranium miners are often exposed to excessively high levels of radon, hundreds have died of lung cancer, and thousands more have had their lives shortened through work-related incidents. According to reports by the International Commission on Radiological Protection, work-related deaths for uranium mining amount to 5,500 to 37,500 per million workers per year, compared to 110 deaths for general manufacturing and 164 deaths for the construction industry.[69] Even more worrying is the evidence that there may be no "safe" level of exposure to the radionuclides at uranium mines. One longitudinal medical study found that low doses of radiation, spread over a number of years, are just as "dangerous" as acute exposure.[70]

In the United States, a 1995 Presidential Advisory Committee on Human Radiation Experiments argued that "the federal government had wronged uranium miners by allowing them to be unwittingly exposed to radiation hazards, and by studying the health effects of their exposures without adequate consent and disclosure." As a result, the government has so far paid $1.3 billion in 28,000 claims to thousands of miners and their families living "downwind" of nuclear test sites and uranium mines. To put these figures in perspective, Navajo uranium miners have a risk of developing lung cancer 28 times greater than Navajos not exposed to uranium mining.[71]

The case of the Shiprock facility in New Mexico illustrates how extremely lethal uranium mining can be. Of the 150 miners working at the mine, 38 have since died of radiation-induced cancer and another 95 have unusual serious respiratory ailments and cancers (meaning that 89 percent of miners, on aggregate, contracted chronic illnesses). That facility, once closed, left 70 acres of raw untreated tailings almost as radioactive as the ore itself. Other studies have shown higher rates of miscarriages, cleft palates, and birth defects among communities living near uranium mines, to say nothing of the psychological damage and guilt miners feel for infecting their families and loved ones with radioactive particles and consequent illnesses. One study recently argued that uranium mining creates "a health crisis of epidemic proportions" when carried out near communities.[72]

A second hazard relates to the radioactive waste mines create. To supply even a fraction of the power stations the industry expects to be online worldwide in 2020 would mean generating millions of metric tons of toxic radioactive tailings every single year. These tailings contain uranium, thorium, radium, and polonium, and emit radon-222. Quite simply, uranium mining results in "the unavoidable radioactive contamination of the environment by solid, liquid, and gaseous wastes."[73]

In developing countries and emerging economies, the impacts from uranium mining can be even more severe, since such governments often lack strong institutional capacity to enforce environmental regulations and statutes. In Africa, for example, the legacy of uranium mining is terrible health, water contamination, and egregious levels of pollution.[74] In Niger, the extraction of uranium and open-pit mining by the former French conglomerate Areva "wreaked havoc" on the nomadic Tuareg peoples, with mining efforts proceeding without any consent whatsoever from local people.[75] In Tanzania, pressure from global uranium mining companies have convinced national planners to open up uranium mines within the Seleous Game Reserve, one of the largest remaining natural habitats in the world for elephants, black rhinos, cheetahs, giraffes, hippos, and crocodiles.[76] The mining companies plan to earn at least $200 million per year from the new Tanzanian mines; the government, by contrast, will receive $5 million.

Uranium mining raises serious questions regarding the equitable treatment of indigenous people, as 70 percent of uranium deposits throughout the world are located on indigenous people's lands.[77] A utilitarian cost-benefit analysis that places more value on the benefits to the many that will enjoy the electricity generated by a nuclear power plant than on the costs to the indigenous people who will lose their land, their health, and their identity violates the prohibitive principle of energy justice. In matters of justice, it is not a question of numbers, but of principle. Any decisions regarding uranium located on indigenous land should involve the full participation of the local inhabitants.

The legacy of nuclear waste

The storage of nuclear waste is an externality because it imposes severe costs on future generations. Nuclear reactors create more than 100 dangerously radioactive chemicals, including Strontium-90, Iodine-131, and Cesium-137, the same toxins found in the fallout from nuclear weapons. Some of these contaminants, such as Strontium-90, remain radioactive for 600 years, concentrate in the food chain, are tasteless, odorless, and invisible, and have been found in the teeth of babies living near nuclear facilities. Strontium-90 mimics milk as it enters the body and concentrates in bones and lactating breasts to cause bone cancer, leukemia, and breast cancer. Babies and children are 10 to 20 times more susceptible to its carcinogenic effects than adults.[78] Plutonium is so dangerous that one pound evenly distributed could cause cancer in every person on earth; it remains radioactive for 500,000 years.[79] It enters through the lungs and mimics iron in the body, migrating to bones, where it can induce bone cancer or leukemia, and to the liver, where it can cause primary liver cancer. It crosses the placenta into the embryo and, like the drug thalidomide, causes gross birth deformities, and it also has a "predilection for the testicles, where it induces genetic mutations in the sperm of humans and other animals that are passed on from generation to generation."[80]

Because nothing is burned or oxidized during the fission process, nuclear plants convert almost all of their fuel to waste with little reduction in mass. Typically, a single nuclear reactor will consume an average of 32,000 fuel rods over the course of its lifetime, and it will also produce twenty to thirty tons of spent nuclear fuel per year – an average of about 2,200 metric tons annually for the entire U.S. nuclear fleet.[81] The global nuclear fleet creates about 10,000 metric tons of high-level spent nuclear fuel each year. About 85 percent of this waste is not reprocessed, and most of it is stored onsite in special facilities at nuclear power plants.

In the United States, spent nuclear fuel is currently stored at 77 different sites. This total includes 63 sites with licensed operating commercial nuclear power reactors, 4 DOE-operated sites, 9 decommissioned reactors, and a proposed reprocessing plant in Morris, Illinois.[82] According to the Congressional Research Service, the total amount of commercial spent fuel was 67,450 metric tons in May 2012.[83] If you include the 2,458 metric tons of U.S. Department of Energy (DOE) spent fuel and high-level waste that was destined for Yucca Mountain, the total amount of waste requiring storage already equals the 70,000 metric-ton limit for the repository imposed by the Nuclear Waste Policy Act of 1982.[84] According to the Blue Ribbon Commission, the amount of waste generated by the industry could potentially increase to between 150,000 and 200,000 metric tons by mid-century.[85]

France is also running out of storage space, and existing sites will likely be full by 2015. A law from 1991 requiring the creation of a geologic storage facility underground was never implemented due to public opposition.[86] In South Korea, an underground repository for the permanent disposal of spent nuclear fuel will not be ready until 2041, but interim storage pools will likely reach maximum capacity by 2024.[87] All the onsite research for the permanent waste repositories in Finland and Sweden has been conducted by the companies themselves with no independent review; and the bedrock in both sites is believed to be less stable and full of more cracks than originally believed, with new evidence revealing that copper canisters could be corroded at the site within a century.[88] As one study concluded, "the management and disposal of irradiated fuel from nuclear power reactors is an issue that burdens all nations that have nuclear power programs. None has implemented a permanent solution to the problem of disposing of high-level nuclear waste, and many are wrestling with solutions to the short-term problem of where to put the spent, or irradiated, fuel as their cooling pools fill."[89]

The issue of waste provoked the nuclear physicist Alvin M. Weinberg to conclude that nuclear power was a "Faustian bargain." Nuclear power creates an unbreakable commitment whereby society benefits from electricity but in return has to bear the costs of managing nuclear waste for astonishingly long periods of time. Plutonium, not found in nature and a byproduct of nuclear power generation, takes 240,000 years to become stable. Stonehenge is slightly more than 4,000 years old; *Homo sapiens* migrated to Europe and Australia 40,000 years ago; and anatomically modern humans have been on earth for only 230,000 years.[90] With nuclear energy, society stakes everything on "the remarkable belief that it can devise social institutions that are stable for periods equivalent to geologic ages."[91] In other words, nuclear reactors will produce waste that will persist longer than our civilization has practiced Catholicism, longer than humans have cultivated crops, and longer than our species has existed.

Air pollution and respiratory health

Power plants and conventional automobiles release a variety of noxious pollutants into the air that threaten human health and the vitality of crops, fisheries, forests, and natural habitats. Again, these externalities have repercussions for both present and future generations. While many of the impacts on human health would quickly disappear if the pollutants were removed from the air, the effect of some of these pollutants on natural ecosystems, fish stocks, and forests will last for many decades.

In the United States, coal-burning power plants pose serious threats to human health. They release an average of 68 percent of their waste by volume directly into the environment, and more than one-third of the coal-fired power plants currently operating in the country (approximately 123 GW out of 300 GW) do not have advanced pollution controls installed. Incidences of drinking wells and surface water being contaminated by leaching from coal waste ponds has swelled, with 67 coal or oil ash waste sites contaminated with heavy metals and other toxic materials. A recent report by the National Research Council examined the damage caused by pollution from energy production and consumption in the United States.[92] The study committee concluded that these damages totaled $120 billion in 2005, excluding any costs of climate change, the effects of mercury, the impacts on ecosystems, and other damages difficult to monetize. The total costs were dominated by sickness and suffering from air pollution associated with electricity generation and vehicle transportation.[93] A 2011 follow-up assessment from the Union of Concerned Scientists estimated at least $100 billion per year in air pollution costs from coal-fired power plants in the United States.[94]

In all likelihood, the true costs are much, much greater. In a national survey of air quality, the American Lung Association warned that 81 million people live in areas of the United States with unhealthy short-term levels of air pollution. Sixty-six million of these Americans live in areas with unhealthy year-round levels of particle pollution, 136 million live in areas with unhealthy levels of ozone, and 46 million live in counties with all three forms of pollution.[95] Lung diseases now affect more than 10 percent of the population and are the third leading cause of death for Americans. The total annual cost for lung disease is estimated to exceed $80 billion in health care expenditures.[96] Nationwide, the EPA has designated 474 counties "non-attainment areas" for unsafe levels of ozone and 224 counties as unsafe areas for fine particulate matter.[97] That is, these counties are so degraded that no additional pollution is permitted under federal law. These areas include all the eastern states from Maine south to Georgia as well as Alabama, Arizona, California, Colorado, Illinois, Indiana, Kentucky, Louisiana, Michigan, Missouri, Nevada, Ohio, Tennessee, Texas, West Virginia, and Wisconsin. It should come as no surprise, then, that at the high end of the range, the health costs related to power plant pollution are estimated to approach $700 billion each year.[98]

Of course, not everything can be quantified monetarily. Researchers at the Harvard School of Public Health estimated that the air pollution from conventional energy sources kills between 50,000 and 70,000 Americans every year.[99] In Texas, 100 different sources (mostly refineries and oil wells), each located within two miles of a school, emit nearly two-thirds of the pollution from all of the state's facilities.[100] Such proximity matters, as the maximum health impacts from emission sources are typically observed closest to the sources.

Children are particularly vulnerable to such pollution. Young children spend more time outdoors, and they breathe 50 percent more air per pound of body weight than adults. Since they breathe polluted air when their respiratory systems are still developing, they become especially susceptible to chronic and life-lasting damage. Exposure to fine particles is associated with many childhood illnesses, and children are less likely to recognize symptoms, leading to delays in treatment and worsening possible damage. Study after study has shown that high periods of air pollution are correlated with a 26 percent increase in sudden infant death syndrome; that infants in high pollution areas are 40 percent more likely to die from respiratory causes; and that, though children make up roughly 25 percent of the population, they comprise almost half of all asthma cases.[101] The EPA has calculated that the number of deaths related to asthma in children has tripled between 1979 and 1996, and that asthma is the third-ranking cause of hospitalization among children under the age of 15. Asthma also accounts for one-third of all emergency room visits by children and is the fourth most common cause of visits by children to the doctor's office.[102]

Sulfur dioxide (SO_2) and nitrogen oxides (NO_x) from power plants are also responsible for acid precipitation that damages forests and lakes. One of the most tragic aspects of this acid deposition is its irreversibility. Once polluted, streams, rivers, and habitats seldom recover. When locations become acidified, the diversity of invertebrates sharply declines, habitats fall apart, densities of salmon and trout fall, and bird breeding is severely impaired. Scientists have found a complicated, nonlinear relationship between continued emissions and polluted areas. In Europe, industry reduced emissions of SO_2 by roughly 80 percent between 1977 and 2007, and sulfur deposition declined in many places by more than 50 percent. However, ecologists found that the mean pH in acidified streams still increased by 0.3 to 0.4 units every 10 years. The basic lesson is that, once polluted, streams take a long time to recover, and even light acidification can completely impair recovery of these fragile ecosystems. Many damaged forests, ecosystems, and water systems, moreover, are often hundreds of miles away from the nearest large power plant, demonstrating how far-reaching and widely dispersed acid rain pollution can be.[103]

In the United States, despite the immense progress made under the Clean Air Act Amendments of 1990, the EPA warns that surface water sulfate concentrations have actually increased in the Blue Ridge provinces of Virginia and that some parts of the Northern Appalachian Plateau region continue to experience dangerously high levels of stream acidification.[104] Acidification has been so severe in central Ontario and Quebec that streams and rivers have not responded to reductions in sulfate deposition. About 95,000 lakes and streams remain damaged and Atlantic Salmon have disappeared in 14 rivers in Nova Scotia and are severely threatened in another 20. New York State, the Adirondacks, and the Catskill mountains are the most sensitive regions to acid inputs in the United States. Recent reductions in SO_2 emissions in these regions have neither improved water quality nor increased the capacity of rivers and streams to neutralize acids. Nearly 25 percent of surveyed lakes in the Adirondacks no longer support any fish. In Western Pennsylvania, acid deposition has been linked with the deterioration of tree health and excessive mortality of mature sugar maples and red oaks. In Virginia, a 13-year trout stream study estimated that even reductions of 40 to 50 percent beyond existing Clean Air Act levels will not support recovery of chronically acidic streams, and will not halt the destruction of streams that are currently suffering worsening levels of acidification. In the Great Smoky Mountains National Park in Tennessee and North Carolina, chronic loading of sulfate and nitrate has made forest sites in the region so vulnerable that scientists estimate that within 80 to 150 years soil calcium reserves will be inadequate to support the growth of healthy trees and merchantable timber.[105]

Acid rain and ozone are not just bad for the environment. During years with hot summers, rates of ozone generation increase and numerous areas with a combined population of more than 135 million people exceed ozone nonattainment levels designed to ensure air quality. During these events, studies have linked short-term ozone exposure to hospital admissions and doctor visits for asthma and other respiratory problems. A 1996 Harvard School of Public Health study found that exposure to ozone was linked to 10,000 to 15,000 hospital admissions and between 30,000 to 50,000 emergency room visits in 13 U.S. cities.[106] Studies have also found the inverse to be true. Health data from 5,000 individuals living in Los Angeles, California confirmed that a 1 percent reduction in ozone engendered substantial benefits for human health with fewer deaths, fewer emergency room visits, and roughly 23,000 avoided instances of respiratory-related diseases.[107]

Power plants are also responsible for almost half of worldwide mercury emissions, with some facilities emitting close to one ton of mercury every year. American coal-fired power plants, for example, release about 50 tons of mercury into the nation's air. The greatest concentrations are found in the southern Great Lakes and Ohio River valley, the Northeast, and scattered areas in the South, with the most elevated concentrations in the Miami and Tampa Bay areas.[108] A comprehensive EPA study on mercury noted that epidemics of mercury poisoning following high doses in Japan and Iraq have demonstrated that neurotoxicity is of greatest concern when developing fetuses are exposed it. Dietary mercury is almost completely absorbed into the blood and distributed to all tissues, including the brain; it also readily passes through the placenta to the fetus and fetal brain.[109] Most Americans do not ingest mercury directly, but accumulate small amounts of the poisonous metal through the consumption of fish. In 2003, 43 states had to issue mercury advisories to warn the public to avoid consuming contaminated fish from in-state water sources.[110] The EPA estimates that as many as 3 percent of women of childbearing age eat sufficient amounts of fish to be at risk from mercury exposure. One study projected that 630,000 infants are born with dangerous levels of mercury every year throughout the country.[111]

The connection between power production and air pollution was vividly documented by the August 2003 Northeast blackout, which not only shut off electricity for 50 million people in the United States and Canada, but also stopped the pollution coming from fossil fuel-fired power plants across the Ohio Valley and the Northeast. In effect, the power outage established an inadvertent experiment for gauging the atmosphere's response to the grid's collapse. Twenty-four hours after the blackout, sulfur dioxide concentrations in the Northeast declined 90 percent, particulate matter dropped by 70 percent, and ozone concentrations fell by half.[112] The blackout demonstrated that conventional electricity generation is by far the largest source of air pollutants that harm human health and damage the natural environment.

Vehicles powered by fossil fuels also spew a variety of unhealthy pollutants and particles directly into the air where they become ingested and inhaled by human beings and ecosystems, contributing to hospital admissions, acid rain, and ozone depletion. Consider emissions of particulate matter (PM), not a specific pollutant itself, but instead a mixture of fine particles of harmful pollutants such as soot, acid droplets, and metals. Conventional automobiles are often the largest single human-caused source of PM, and for those places with stringent emissions requirements for vehicles, such as California or the EU, the second largest human source after power plants (forest fires and dust storms are the leading non-human sources). Thousands of medical studies have strongly associated inhalation of PM with heart disease, chronic lung disease, and some forms of cancer.

Table 3.1 Selected causes of death in the United States, 2008

Cause	Estimated annual deaths
Heart Disease	652,000
Cancer	559,300
Stroke	143,600
Alzheimer's Disease	71,600
Particulate Matter Pollution	65,600
Influenza	63,000
Nephritis	43,900
Breast Cancer	40,900
Automobile Fatalities	36,700
Prostate Cancer	30,100
HIV	18,000
Drunk Driving	16,700

Source: U.S. Centers for Disease Control and Prevention and the
National Institutes for Health.

These findings were confirmed, again, when the burning of 650 oil wells
in Kuwait during the 1991 Persian Gulf War affected U.S. soldiers with
respiratory illness, asthma, and emphysema.

Using some of the most recently available data, Table 3.1 shows that in
the United States, deaths from PM pollution are comparable to those from
Alzheimer's disease and influenza and greater than the deaths from nephritis,
septicemia, breast cancer, automobile accidents, prostate cancer, HIV/AIDS,
and drunk driving.[113] In France, the Agency for Health and Environmental
Safety projects that normal automobile emissions kill 9,513 people per year
and result in 6 to 11 percent of all lung cancer cases identified in people above
30 years of age.[114] Another report from the World Health Organization investigating automobile emissions in Austria, Switzerland, and France calculated
40,000 deaths per year from PM emissions from automobiles.[115]

Athens, Milan, and other European cities are today experiencing crumbling
historic monuments and high rates of respiratory diseases. Cities in Mexico,
India, and China are witnessing far worse levels of smog due to declining air
quality standards. Only 1 percent of China's 560 million city dwellers breathe
air considered safe by the European Union; 297 of the largest 300 Chinese cities do not meet the minimal environmental standards for ambient air pollution
set by the United States; and the World Bank estimates that the economic burden of premature mortality associated with air pollution in China was greater
than 1.16 percent of GDP in 2007.[116] Even in the United States, where clean
air regulations have been strict, local topography can create temporary temperature inversions that trap pollution, often in dense urban areas like Denver and
Los Angeles. The greater Los Angeles region, for instance, is home to 16 million automobiles and has the worst air quality in the United States, and these
vehicles contribute up to 89 percent of the average cancer risk from air toxics.[117]

Transportation, which relies almost entirely on petroleum products, accounts for roughly one-third of America's energy consumption and about 40 percent of global energy consumption. Another 40 percent of total energy consumption is used to generate electricity for the industrial, commercial, and residential sectors of society. In this last section we have seen the human costs of these two types of energy use, costs that sadly affect children more than they do adults. Given that these two types of energy use are also responsible for the majority of CO_2 emissions in the United States, there are compelling reasons to find alternatives. It is a wonder that a more fundamental shift has not already occurred, as effective alternatives exist and are only awaiting the necessary investments of capital to become commercially viable. The argument most often given, that it does not make economic sense, is hardly plausible given the aggregated costs associated with the current system, as described above. All that is lacking is the political and social will to make the change, and a proper appreciation for the real (human) costs and benefits.

Conclusion

Global warming and other externalities associated with modern energy systems threaten to compromise the ability of global commons like the atmosphere and the oceans to sustain and nurture a flourishing human life for millions of people throughout the world. If the acidification of the ocean's surface waters, for instance, resulting from the absorption of increased carbon in the atmosphere, were to lead to a significant collapse of species within marine ecosystems, the consequences would be catastrophic for the hundreds of millions of people who rely on fish as their primary source of protein. It is hard to imagine a greater injustice than undermining the basic conditions required for sustaining life – fresh water, food, arable land – and yet for many people in places like Africa, the South Pacific, and South Asia, this is precisely what our energy sector is doing, poisoning the sources and springs of life.

The loudest argument for continuing with the status quo, and the one which is received with most deference by both politicians and the public, is that the economic costs of reducing our reliance on fossil fuels are simply too high: regardless of what happens in the future, for the time being we have no choice. This argument, however, is only plausible so long as we ignore the enormous *present* costs – both market and nonmarket – that are simply not reflected in the prices of fossil fuels, but that we pay nonetheless. It is not really a question of what the cost is, in fact, but who has to pay it. Under the present energy system, it is paid by those who can least afford it: Islanders in Kiribati and Tuvalu faced with losing their homes under the sea; farmers in Africa struggling with drought and salinization; miners dying of lung cancer; children struggling to breathe; communities forced to relocate because their soil and water is contaminated; taxpayers responsible for a stretched health care system that has to cope with the health impacts of polluted air and water; and government agencies (and taxpayers again) charged with cleaning up oil

spills, mine tailings and fly ash spills. These costs are all real, and can be mon-
etized, and have an effect on the economy. If they were included in the price
of energy, it would become clear not only that we can afford to wean ourselves
from fossil fuels, but that it would actually be profitable to do so. But in that
case, someone else would have to pay the cost of this shift, namely, those who
benefit most from current energy systems.

   In the list of bullet points that follows, we quickly recap the major externalities associated with the energy sector that we discuss in this chapter.

- Climate change impacts are not just a future threat but a present reality:
  An increased frequency and intensity of extreme weather events, rising
  sea levels, droughts, floods and heat-related stresses have already caused
  billions of dollars worth of damage and resulted in the deaths of hundreds of thousands of people.
- Greenhouse gas emissions from the global energy sector are not being
  reduced; on the contrary, they are increasing. In the spring of 2013, the
  atmospheric concentration of carbon dioxide reached 400 ppm.
- At present, climate change has a disproportionate impact on the most
  vulnerable populations of the world, people who are least able to adapt
  to changing conditions, and have done the least to contribute to the
  problem. If any of the various tipping points are surpassed, the damages
  for the entire globe could be catastrophic.
- The economic costs of climate change are already large: Over the past
  15 years, economic damages from extreme weather events have amounted
  to approximately $1 trillion.
- The various extraction processes for fossil fuels (coal, oil, and gas) have
  both immediate and lasting environmental costs (with ramifications for
  human health), including: the dumping of highly toxic "produced water"
  from off-shore drilling platforms; the release of toxic chemicals through
  the discharge of drilling muds and cuttings; damage of marine habi-
  tat from oil spills; chemical and hydrologic alterations of streams from
  mountaintop mining; contamination of streams, rivers, and groundwater
  from effluent discharges associated with coal mines and ecological degradation of land from stored fly ash.
- Uranium mining annually generates millions of tons of toxic radioactive
  tailings that contain uranium, thorium, radium, and polonium, and emit
  radon-222; uranium mines release harmful levels of gamma radiation;
  for miners and local communities, exposure to radiation hazards leads to
  increased risk of cancer, birth defects, cleft palates, and miscarriages.
- The storage problem for spent nuclear fuel from nuclear reactors has
  not yet been solved, and most countries are now running out of space
  in existing facilities; some of the contaminants created by nuclear reactors remain radioactive for more than 600 years, concentrate in the food
  chain, and have deleterious health impacts for human beings; plutonium
  remains radioactive for 500,000 years.

- Power plants and automobiles release a variety of pollutants into the air through the combustion of fossil fuels, with immediate and lasting environmental and human health costs; these include mercury poisoning, asthma, acid precipitation, and respiratory-related diseases, among other impacts.
- The economic costs of the damages from energy production and consumption are huge; in 2005, U.S. damages totaled $120 billion, excluding damages from climate change, the effects of mercury, ecological degradation, and other costs difficult to monetize.

Notes

1. Nicholas Stern, *The Economics of Climate Change: The Stern Review* (Cambridge: Cambridge University Press, 2007); for the controversy, see, e.g., William D. Nordhaus, "A Review of the Stern Review on the Economics of Climate Change," *Journal of Economic Literature* 55 (2007): 686–702; Partha Dasgupta, "Comments on the Stern Review's Economics of Climate Change," rev. (Cambridge: University of Cambridge, 2012); Hal R. Varian, "Recalculating the Costs of Global Climate Change," *The New York Times,* December 14, 2006.
2. William D. Nordhaus, "'The Question of Global Warming': An Exchange," *The New York Review of Books,* September 25, 2008.
3. Dimitri Zenghelis, "'The Question of Global Warming': An Exchange," *The New York Review of Books,* September 25, 2008 (quoting Freeman Dyson, "The Question of Global Warming").
4. See Donella Meadows, Jorgen Randers, and Dennis Meadows, *Limits to Growth: The 30-Year Update* (London: Earthscan, 2005); see also Herman E. Daly, *Beyond Growth* (Boston, MA: Beacon Press, 1996); *Valuing the Earth: Economics, Ecology, Ethics,* ed. Herman E. Daly and Kenneth N. Townsend (Cambridge, MA: MIT Press, 1993); Richard Heinberg, *The End of Growth: Adapting to Our New Economic Reality* (Gabriola Island, BC: Clairview/New Society Publishers, 2011).
5. See Heinberg, *The End of Growth.*
6. Alvin M. Weinberg, "Immortal Energy Systems and Intergenerational Justice," *Energy Policy* (February, 1985): 51–59.
7. UNESCAP, *Low Carbon Green Growth Roadmap for Asia and the Pacific* (Bangkok: UNESCAP, 2012).
8. Tim Jackson, "Renewable Energy: Great Hope or False Promise?" *Energy Policy* (January/February, 1991): 7.
9. See David Humphreys, *Logjam: Deforestation and the Crisis of Global Governance* (London: Earthscan, 2006), 216.
10. Matthew Bishop, *The Essential Guide to Economics* (Bloomberg Press, 2009), 34. See also Anthony D. Owen, "Environmental Externalities, Market Distortions, and the Economics of Renewable Energy Technologies," *The Energy Journal* 25, no. 3 (2004): 127–152.
11. U.S. Energy Information Administration, *Electricity Generation and Environmental Externalities: Case Studies* (Washington, DC: U.S. Department of Energy, 1995).

12. Noah D. Hall, "Political Externalities, Federalism, and a Proposal for an Inter-state Environmental Impact Assessment Policy," *Harvard Environmental Law Review* 32 (2007/2008): 50–94.

13. David E. Nye, *America as Second Creation: Technology and Narratives of New Beginnings* (Cambridge, MA: MIT Press, 2003), 192–193.

14. Gavin Bridge, "Contested Terrain: Mining and the Environment," *Annual Review of Environment and Resources* 29 (2004): 205–259.

15. William J. Baumol and Wallace E. Oates, "Externalities: Definition, Significant Types, and Optimal-Pricing Conditions," in *The Theory of Environmental Policy* (Cambridge, MA: Cambridge University Press, 1988), 14–35.

16. Steven G. Medema, "Mill, Sidgwick, and the Evolution of the Theory of Market Failure," working paper, University of Colorado Department of Economics, July 2004.

17. See Alfred Marshall, *Principles of Economics: An Introductory Volume* (New York: The Macmillan Company, 1890 [1920]) and A. C. Pigou, *Wealth and Welfare* (London: MacMillan and Company, 1912), later to become his *The Economics of Welfare* (London: MacMillan and Company, 1920 [1924]).

18. See Garrett Hardin, *The Tragedy of the Commons,* 162 SCI 1243–1248 (1968).

19. See ibid., 1245.

20. Richard S. J. Tol, "The Economic Effects of Climate Change," *Journal of Economic Perspectives* 23, no. 2 (2009): 29–51.

21. Intergovernmental Panel on Climate Change (IPCC), *Climate Change 2007: Synthesis Report* (Geneva: IPCC, 2008), 36.

22. United States Environmental Protection Agency, *Inventory of U.S. Greenhouse Gas Emissions and Sinks, 1990 to 2010* (Washington, DC: U.S. EPA, 2012).

23. R. Sobin, "Energy Myth Seven: Renewable Energy Systems Could Never Meet Growing Electricity Demand in America," in *Energy and American Society: Thirteen Myths,* ed. B. K. Sovacool and M. A. Brown (New York: Springer, 2007), 171–200.

24. David G. Hawkins, Daniel A. Lashof, and Robert H. Williams, "What to Do About Coal?" *Scientific American* (September 2006): 69–72.

25. David J. C. MacKay, *Sustainable Energy: Without the Hot Air* (UIT Cambridge, 2009), 35.

26. Richard Lovett, "Natural Gas Greenhouse Emissions Study Draws Fire," *Nature,* April 15, 2011, 242.

27. F. O'Sullivan and S. Paltsev, "Shale Gas Production: Potential Versus Actual Greenhouse Gas Emissions," *Environmental Research Letters* 7, no. 4 (2012).

28. Robert W. Howarth, Renee Santoro, and Anthony Ingraffea, "Methane and the Greenhouse Gas Footprint of Natural Gas from Shale Formations," *Climatic Change* (2011).

29. B. K. Sovacool, "Valuing the Greenhouse Gas Emissions from Nuclear Power: A Critical Survey," *Energy Policy* 36, no. 8 (2008): 2940–2953.

30. F. Barnaby and J. Kemp, *Secure Energy? Civil Nuclear Power, Security, and Global Warming* (Oxford: Oxford Research Group, 2007).

31. Eileen Claussen and Janet Peace, "Energy Myth Twelve: Climate Policy Will Bankrupt the U.S. Economy," in *Energy and American Society,* ed. B. K. Sovacool and M. A. Brown (New York: Springer, 2007), 311–340.

32. Stern, *The Economics of Climate Change;* IPCC, "Summary for Policymakers."

33. Heather Stewart and Larry Elliot, "Nicholas Stern: 'I Got it Wrong on Climate Change – it's Far, Far Worse,'" *The Observer,* January 26, 2013.
34. B. Sudhakara Reddy and Gaudenz B. Assenza, "The Great Climate Debate," *Energy Policy* 37 (2009): 2997–3008.
35. B. K. Sovacool, "Conceptualizing Hard and Soft Paths for Climate Change Adaptation," *Climate Policy* 11, no. 4 (2011): 1177–1183; B. K. Sovacool, "Perceptions of Climate Change Risks and Resilient Island Planning in the Maldives," *Mitigation and Adaptation of Strategies for Global Change* (2012); B. K. Sovacool, "Expert Views of Climate Change Adaptation in the Maldives," *Climatic Change* (2012).
36. A. Rawlani and B. K. Sovacool, "Building Responsiveness to Climate Change through Community Based Adaptation in Bangladesh," *Mitigation and Adaptation Strategies for Global Change* 16, no. 8 (2011): 845–863.
37. World Health Organization, *Climate Change and Human Health* (Geneva: WHO, 2003).
38. Royal United Services Institute (RUSI), *Socioeconomic and Security Implications of Climate Change in China* (Washington, DC: CNA, 2009).
39. CNA, Climate Change, State Resilience, and Global Security Conference, CNA Conference Center, Alexandria, Virginia, November 4, 2009.
40. David M. Catarious and Ralph E. Espach, *Impacts of Climate Change on Columbia's National and Regional Energy Security* (Washington, DC, 2009).
41. B. K. Sovacool, "Conceptualizing Hard and Soft Paths for Climate Change Adaptation," *Climate Policy* 11, no. 4 (2011): 1177–1183; Sovacool, "Perceptions of Climate Change Risks"; Sovacool, "Expert Views of Climate Change."
42. Augusto De La Torre, Pablo Fajnzybler, and John Nash, *Low-Carbon Development: Latin American Responses to Climate Change* (Washington, DC: World Bank Group, 2010).
43. H. Meenawat and B. K. Sovacool, "Improving Adaptive Capacity and Resilience in Bhutan," *Mitigation and Adaptation Strategies for Global Change* 16, no. 5 (2011): 515–533.
44. M. A. Brown and B. K. Sovacool, *Climate Change and Global Energy Security: Technology and Policy Options* (Cambridge, MA: MIT Press, 2011).
45. Ruth Gordon, "Climate Change and the Poorest Nations: Further Reflections on Global Inequality," *University of Colorado Law Review* (2007): 1559–1624.
46. David Waskow and Carol Welch, "The Environmental, Social, and Human Rights Impacts of Oil Development," in *Covering Oil: A Reporter's Guide to Energy and Development,* ed. Svetlana Tsalik and Anya Schiffrin (New York: Open Society Institute, 2005), 101–123.
47. Ibid.
48. Oil and gas are themselves toxic. Both contain significant quantities of hydrogen sulfide, a substance that is potentially fatal and extremely corrosive to equipment such as drills and pipelines.
49. Waskow and Welch, "Environmental, Social, and Human Rights Impacts."
50. Ibid.
51. Charles H. Peterson et al., "Ecological Consequences of Environmental Perturbations Associated with Offshore Hydrocarbon Production: A Perspective on the Long-Term Exposures in the Gulf of Mexico," *Canadian Journal of Fisheries and Aquaculture Science* 53 (1996): 2637–2654.

52. Waskow and Welch, "Environmental, Social, and Human Rights Impacts."
53. Patricio Silva, "National Energy Policy," hearing before the House Subcommittee on Energy and Air Quality, February 18, 2001, 113–116.
54. Sobin, "Energy Myth Seven."
55. Emily S. Bernhardt and Margaret A. Palmer, "The Environmental Costs of Mountaintop Mining Valley Fill Operations for Aquatic Ecosystems of the Central Appalachians," *Annals of the New York Academy of Sciences* 1223 (2001): 39–57.
56. M. A. Palmer et al., "Mountaintop Mining Consequences," *Science* 327 (2010): 148–149.
57. Quoted in Natural Resources Defense Council, "Mountain Top Removal Coal Mining," June 23, 2011.
58. "Mining and Pollution: National Treasure," *The Economist* 387, no. 8582 (2008): 47.
59. Ananth P. Chikkatur, Ankur Chaudhary, and Ambuj D. Sagar, "Coal Power Impacts, Technology, and Policy: Connecting the Dots," *Annual Review of Environment and Resources* 36 (2011): 101–138.
60. United Nations Environment Program, UNDP, NATO, and OSCE, *Mining for Closure: Policies and Guidelines for Sustainable Mining Practice and Closure of Mines* (Norwegian Ministry of Foreign Affairs, 2005).
61. Adriana Eftimie, Katherine Heller, and John Strongman, *Gender Dimensions of the Extractive Industries: Mining for Equity,* Washington, DC, World Bank, Extractive Industries and Development Series #8, August 2009.
62. Daniel M Franks, David Brereton, and Chris J Moran, "Managing the Cumulative Impacts of Coal Mining on Regional Communities and Environments in Australia," *Impact Assessment and Project Appraisal* 28, no. 4 (2010): 299–312.
63. Yun Zhou, "Why Is China Going Nuclear?" *Energy Policy* 38 (2010): 3755–3762.
64. Virginia Sharpe, "Clean Nuclear Energy? Global Warming, Public Health, and Justice," *Hastings Center Report* 38, no. 4 (2008): 16–18.
65. David Thorpe, "Extracting Disaster," *The Guardian,* December 5, 2008.
66. V. N. Mosinets, "Radioactive Wastes from Uranium Mining Enterprises and Their Environmental Effects," *Atomic Energy* 70, no. 5 (1991): 348–354.
67. V. V. Shatalov, M. I. Fazlullin, R. I. Romashkevich, R. N. Smirnova, and G. M. Adosik, "Ecological Safety of Underground Leaching of Uranium," *Atomic Energy* 91, no. 6 (2001): 1009–1015.
68. G. M. Mudd, "Uranium Mining in Australia: Environmental Impact, Radiation Releases and Rehabilitation," in *Protection of the Environment from Ionizing Radiation: The Development and Application of a System of Radiation Protection for the Environment,* ed. IAEA (Vienna: International Atomic Energy Agency, 2003), 179–189.
69. Roxby Action Collective and Friends of the Earth, *Uranium Mining: How It Affects You* (Sydney: Friends of the Earth, 2004).
70. Ibid.
71. Sharpe, "Clean Nuclear Energy?" 16–18.
72. Barbara Rose Johnston, Susan E. Dawson, and Gary E. Madsen, "Uranium Mining and Milling: Navajo Experiences in the American Southwest," in *The Energy Reader,* ed. Laura Nader (London: Wiley-Blackwell, 2010), 132–146.
73. Mosinets, "Radioactive Wastes from Uranium Mining," 348.

74. Thorpe, "Extracting Disaster."
75. Sharpe, "Clean Nuclear Energy?" 16–18.
76. "Tanzania 'Will Mine Uranium on Selous Game Reserve,'" *BBC News,* July 1, 2011, http://www.bbc.co.uk/news/world-africa-13989264.
77. Roxby Action Collective and Friends of the Earth, *Uranium Mining.*
78. Helen Caldicott, "Nuclear Power Isn't Clean, It's Dangerous," *Sydney Morning Herald,* August 27, 2001.
79. Hellen Caldicott, *Nuclear Madness* (New York: Norton and Norton, 1994).
80. Caldicott, "Nuclear Power Isn't Clean."
81. Ibid.
82. John Werner, Cong. Research Serv., R42513, U.S. Spent Nuclear Fuel Storage 17 (2012).
83. Ibid., 5.
84. Ibid.
85. Blue Ribbon Commission on America's Nuclear Future, Report to the Secretary of Energy, January 2012, 14.
86. Yves Marignac, Benjamin Dessus, Helene Gassin, and Bernard Laponche, *Nuclear Power: The Great Illusion* (Paris: Global Chance, 2008).
87. Chang Min Lee and Kun-Jai Lee, "A Study on Operation Time Periods of Spent Fuel Interim Storage Facilities in South Korea," *Progress in Nuclear Energy* 49 (2007): 323–333.
88. "Further Nuclear Reactor Construction Delays," *Helsingin Sanomat,* August 11, 2007, 11.
89. Allison Macfarlane, "Interim Storage of Spent Fuel in the United States," *Annual Review of Energy and Environment* 26 (2001): 201–235.
90. Greenpeace, *Nuclear Power: A Dangerous Waste of Time* (Amsterdam: Greenpeace International, 2009).
91. Alvin M. Weinberg, "Social Institutions and Nuclear Energy," *Science* 177, no. 4043 (1972): 27–34.
92. National Research Council (NRC), *Hidden Costs of Energy: Unpriced Consequences of Energy Production and Use* (Washington, DC: The National Academies Press, 2009).
93. Ibid.
94. Barbara Freese, Steve Clemmer, Claudio Martinez, and Alan Nogee, *A Risky Proposition: The Financial Hazards of New Investments in Coal Plants* (Washington, DC: Union of Concerned Scientists, 2011).
95. American Lung Association, *Summary of the American Lung Association's Annual Clean Air Test* (Washington, DC: ALA, 2005).
96. Joseph J. Romm and Christine A. Ervin, *How Energy Policies Affect Public Health* (Washington, DC: Solstice, 2005). Adjusted to 2007.
97. Debra A. Jacobson, "Increasing the Value and Expanding the Market for Renewable Energy and Energy Efficiency with Clean Air Policies," *Environmental Law Review* 37 (2007): 10135–10137.
98. See Mark A. Delucchi, James J. Murphy, and Donald R. McGubbin, "The Health and Visibility Cost of Air Pollution: A Comparison of Estimation Methods," *Journal of Environmental Management* 64 (2002): 139–152; Isabelle Romieu, Jonatham M. Samet, Kirk R. Smith, and Nigel Bruce, "Outdoor Air Pollution and Acute Respiratory Infections among Children in Developing Countries," *Journal of Occupational and Environmental Medicine* 44 (2002):

640–649; K. Katsouyanni and G. Pershagen, "Ambient Air Pollution Exposure to Cancer," *Cancer Causes and Control* 8 (1997): 284–291.

99. Harvard School of Public Health, *Impact of Pollution on Public Health* (Cambridge, MA: Harvard University, 2001).

100. Deanne M. Ottaviano, *Environmental Justice: New Clean Air Act Regulations and the Anticipated Impact on Minority Communities* (New York: Lawyer's Committee for Civil Rights Under Law, 2003).

101. Conrad G. Schneider, *Death, Disease, and Dirty Power: Mortality and Human Health Damage Due to Air Pollution from Power Plants* (Boston: Clean Air Task Force, 2000); Maria T. Padian, *New York's Dirty Street: The Power Plant Pollution Loophole* (New York: Pace Energy Project, 1998); Michael T. Kleinman, *The Health Effects of Air Pollution on Children* (Los Angeles: University of California Press, 2000).

102. U.S. Environmental Protection Agency, *Asthma Facts* (Washington, DC: EPA, 2004).

103. R. A. Kowalik, D. M. Cooper, C. M. Evans, and S. J. Ormerod, "Acid Episodes Retard the Biological Recovery of Upland British Streams from Acidification," *Global Change Biology* 13 (2007): 2439–2452.

104. U.S. Environmental Protection Agency, *Air Pollution Facts* (Washington, DC: EPA, 2003).

105. Armond Cohen, "National Energy Policy: Coal," hearing before the House Committee on Energy and Commerce, March 14, 2001, 65–71.

106. Romm and Ervin, *How Energy Policies Affect Public Health.*

107. Alan J. Krupnick and Winston Harrington, "Ambient Ozone and Health Effects: Evidence from Daily Data," *Journal of Environmental Economics and Management* 18 (1990): 1–18.

108. Sobin, "Energy Myth Seven."

109. U.S. Environmental Protection Agency, *Mercury Study – Report to Congress* (Washington, DC: EPA, 1997); U.S. Environmental Protection Agency, *Fact Sheet on Mercury* (Washington, DC: EPA, 2001).

110. U.S. Environmental Protection Agency, *2003 Mercury Advisory Listing* (Washington, DC: EPA, 2004).

111. Quoted in Carl Pope, "The State of Nature: Our Roof Is Caving In," *Foreign Policy* (July/August 2005): 67–71.

112. T. Lackson et al., "The 2003 North American Electrical Blackout: An Accidental Experiment in Atmospheric Chemistry," *Geophysical Research Letters* 31 (2004): 3106.

113. Benjamin K. Sovacool, "A Transition to Plug-In Hybrid Electric Vehicles (PHEVs): Why Public Health Professionals Must Care," *Journal of Epidemiology and Community Health* 64, no. 3 (2010): 185–187.

114. Julio Godoy, "Auto Emissions Killing Thousands," *Common Dreams News Release,* June 3, 2004, http://www.commondreams.org/headlines04/0603–08.htm.

115. Ibid.

116. Chen Gang, "Energy Efficiency: High Politics in China," presentation at the Conference on Energy Efficiency, Singapore, March 27–28, 2008.

117. Benjamin K. Sovacool, "A Transition to Plug-In Hybrid Electric Vehicles (PHEVs): Why Public Health Professionals Must Care," *Journal of Epidemiology and Community Health* 64, no. 3 (2010): 185–187.

4   The economic dimension

Inequality, poverty, and rising prices

The test of our progress is not whether we add more to the abundance of those who have much; it is whether we provide enough for those who have too little.
– Franklin D. Roosevelt, "One Third of a Nation,"
FDR's Second Inaugural Address, January 20, 1937

Introduction

Coal and petroleum fueled the extraordinary economic growth of the past two centuries. Beginning in the eighteenth century, the development of industrial technology enabled humankind to harness a vast store of solar energy concentrated over millennia. The substitution of this concentrated energy for human and animal labor power literally transformed the face of the planet. More than any other single factor, the ability to exploit what Nicholas Georgescu-Roegen called "mankind's entropic dowry"[1] made possible the explosive growth in productivity that created the wealth and prosperity of modern industrial nations. Today, millions of ordinary people enjoy a level of material luxury that could not have been imagined by even the most privileged classes of pre-industrial societies. Yet President Franklin D. Roosevelt's remark begs us to ask a troubling question: has this impressive global system of energy production and consumption left anybody behind?

When coal and oil were cheap, it was possible to distribute the benefits of modern energy supply systems widely across a large portion of society; the rise of an affluent, broad-based middle class in the past century was based primarily on the availability of cheap energy. However, while fossil-based energy systems *can* be used for the benefit of many people, they also have an enormous potential to magnify economic inequalities. Fossil fuels are scarce and therefore rivalrous goods. Because every aspect of the industrial economy is now directly or indirectly dependent on the energy from fossil fuels, fluctuations in the prices of fossil fuels can have a devastating impact on the poorer sections of the population, limiting access to such basic goods as food, heating, lighting, and transportation. Correspondingly, ownership and control of the fossil fuel energy supply ensure economic dominance. This economic power translates easily into political power, and the end result is an economic system that rewards those with the greatest access to energy services, while marginalizing those who cannot afford such access.

The broad distribution of energy services that accompanied the expansion of the middle classes never really extended beyond the borders of the core industrial nations. Throughout the rest of the world, the vast majority of people had limited or no access to energy for most of the twentieth century. Even today, with the rapid growth of emerging markets like China and India, 1.3 billion people – roughly one out of five – still live without electricity and 2.5 billion average 1,000 kWh/household.[2] This situation, of course, is in part the result of the historical trajectories of different peoples, nations, and continents. Even if we accept that energy justice requires an equitable distribution of energy services within a given nation, it is reasonable to ask whether the same can be said within the international context. There is no injustice involved in the fact that a remote Amazonian tribe is without access to energy. Yet such isolated and disconnected tribes are now the exception. Countries and people around the world are bound together by a global economy that is entirely dependent on fossil fuels. In this globalized world, there are few people who do not feel the impact of fluctuations in the international markets for commodities. An increase in the price of oil affects the price of grain in North Africa; it affects the price of copper in Zambia and the tourist industry in Thailand. In such a world, to have limited or no access to energy services is to be at a profound disadvantage.

As we shall see in this chapter, the inequalities associated with the energy dependence of the global economy will only increase as fossil fuels become more expensive. While there remains some doubt as to when the nonrenewable reserves of fossil fuels will begin running out, all the evidence suggests that the era of cheap fossil fuels is over.[3] The widely shared prosperity that characterized industrial nations in the latter half of the past century, insofar as it was based on an abundance of inexpensive energy, was a one-off show that is now coming to an end. And yet we persist in acting as though the performance can be repeated – as though continued reliance on fossil fuels will continue to bring prosperity for ever-increasing numbers of people. This is an illusion that must be abandoned. An equitable distribution of energy services will only be accomplished by the development of a new global energy system that is based, to the greatest extent possible, on renewable sources of energy and distributed generation.

Energy inequality

Sustainable development and affordable energy

For billions of people within the developing world, gaining reliable and affordable access to energy services is an essential condition for improving their situation. A study by the World Bank, for example, found that villages in Africa that lack electricity are less likely to have a secondary school or a post office, two key institutions for participating effectively in the global economy.[4] The absence of these institutions does not mean that such villages remain unaffected by the globalized economy; on the contrary, throughout the developing world rural communities struggle to adapt local practices and expectations to the growing international market for agricultural goods. Subsistence farming and production for local markets give way to cash crops produced for export, often by large agricultural companies that purchase valuable land from the government, dispossessing smallholders who then find themselves unable to earn enough money to buy food. Education and the chance to develop diversified skills that are more appropriate for the globalized economy can significantly reduce the vulnerability of such communities to the hazards of volatile international commodity markets; yet without energy services of some sort, such opportunities are often lacking.

The reality of energy poverty, which affects almost half the world's population in some way, is a significant factor in the international debate over climate change and how to allocate the burdens of reducing global greenhouse gas emissions. Developing countries quite reasonably contend that they have the right to follow in the footsteps of other industrialized nations by pursuing a path of economic development that depends on the exploitation of fossil fuel energy. While the conventional assumption that there is a direct correlation between economic growth and an increase in energy consumption is no longer accepted unquestioningly (with good reason[5]), there is little doubt that increasing energy consumption has been *one way* historically to fuel economic growth. For the past 20 years the economic trajectory of much of Southeast Asia has been quite remarkable. Figure 4.1 shows how an increase in electricity production in Vietnam has been tracked consistently by a growth in GDP.

Indonesia, a much larger economy, shows much the same story. Between 1991 and 2010, its GDP grew from $128 billion to $706 billion. In the same period, its annual gross production of electricity increased from just fewer than 40,000 GWh to approximately 160,000 GWh. Of course, because much of this electricity is generated by burning coal, the carbon dioxide emissions from such countries are rapidly increasing.

In response to calls from the international community to curb the rise in greenhouse gas emissions, developing countries sometimes argue that the present needs of addressing poverty through economic growth outweigh the future hazards of a warming climate. In the context of energy justice, this situation presents a dilemma that resists any easy resolution. On the one hand, investing in energy infrastructure in the developing world will undoubtedly help to alleviate the present suffering of millions of people from poverty, malnutrition, unemployment, poor sanitation, and inadequate health care services. On the other hand, according to the most recent projections of the International Energy Agency, if nothing changes in current investment trends in the energy sector, especially in developing countries, within a few years we

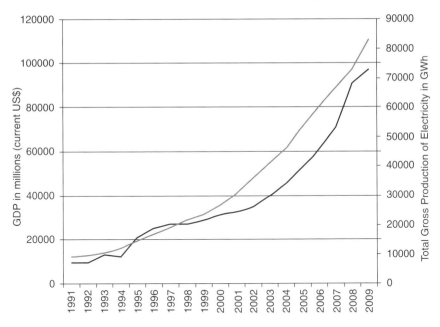

Figure 4.1 Time line showing annual growth in GDP and electricity production in Vietnam, 1991 to 2010

Source: World Bank and International Energy Agency.

will have gone beyond the threshold where we can plausibly keep global mean temperatures from rising 2° Celsius.[6]

We have already discussed in Chapter 3 the fallacy regarding climate change impacts as an issue that primarily concerns future generations, so we leave that argument to one side. In this chapter, our focus is on the economic dimension of energy justice and inequalities in the distribution of energy services. The question is whether there is any reason to forego the kind of development in energy infrastructure we have seen for the past few decades, purely on the basis of economic considerations. Energy inequality exists not only among countries but also within countries, both in the developed and developing worlds. In fact, it cuts across the neat division between the global North and South that presents such an obstacle to progress in the international climate change regime. Statistics describing a nation's gross production of electricity tells us very little about the distribution of energy services within the country, whether it is Indonesia or the United Kingdom. As we will argue more fully in subsequent chapters, investing in the construction of large-scale centralized fossil fuel-based energy infrastructure, rather than diminishing energy inequality in the developing world, in many cases exacerbates it. Growth of GDP is not the only way of measuring the wealth of a country, and it completely ignores the question of distribution; the massive energy projects currently underway from South Africa to India tend to create wealth for a minority while further marginalizing the majority.

A factor of central importance in this context is that the prices of fossil fuels are rising, and according to most projections will continue to do so (see the second part of this chapter for details), driven by high costs of extraction and burgeoning global demand. Only those who can afford it have access to the energy derived from fossil fuels, and increasingly, this will exclude a large portion of the global population.

Rising oil prices, for instance, are already affecting the poor in the United States. For many U.S. residents, not owning a car can make it difficult to obtain work. The necessity of commuting to work by car can consume a large portion of annual income for people in the lower income brackets. In 2012, the average annual cost of operating a car – excluding its price tag – encompassing gasoline, maintenance, insurance, and financing, surpassed $8,900, or about 60 cents per mile. Vehicle ownership costs rose 3.4 percent from 2010 to 2011, and from 2011 to 2012 the cost of fuel rose 14.8 percent.[7] Responding to these trends, advocates for social justice sometimes argue for increased oil production, pointing to the disproportionate impact of high fuel prices on the poor. Yet per capita energy consumption in the United States is already among the highest in the world.[8] If the United States, with its numerous subsidies for fossil fuels, and its substantial domestic production, is unable to maintain gasoline prices at a level that is fair for the poor, there is little chance that developing countries will manage to provide an equitable distribution of energy services.

Given these considerations, we have to ask whether further development of conventional energy systems makes sense even from the perspective of an equitable distribution of economic opportunities (leaving aside the issue of climate change). Of course, many people in China and India and elsewhere have benefited from the construction of fossil fuel-based energy systems (and the manufacturing industries that rely on them). But many have not, and the situation of some has been considerably worsened. More energy is not a solution to energy inequality when the type of energy provided puts it out of the price range of those who are most in need of it. It is not only a question of more, but what kind of energy, how it is distributed, and to whom. As long as the global economy remains dependent on fossil fuels, the poorest people in developed countries will continue to pay a disproportionate share of their income to keep their homes warm and their cars running, and even poorer people in the developing world will continue to struggle with rising food prices and a lack of basic services.

Unequal distribution of energy services

Energy is a prerequisite for almost all economic activity today. This means that access to energy not only reflects economic and social inequalities, but given the scarcity of fossil fuel resources, can also reinforce and magnify such inequalities. Consider a seemingly trivial yet illustrative example: when muscle power was the primary energy used for moving from one place to another, the difference between a peasant and the lord of the manor with respect to available energy was no more than the difference between the muscle power of a human being and that of a horse (and even then, it might have been possible for the peasant to acquire the use of a horse). By contrast, compare the travel options of an employee just making ends meet at a grocery store in Brooklyn with those of a wealthy professional working at an accounting firm in Manhattan. The former can purchase a bus ticket to visit a town in the next state; the latter can board a plane to fly across the ocean for a weekend game of golf. Three orders of magnitude separate the power at the disposal of the accountant (approximately 120 MW for a modern jet) from the power available to the employee at the grocery store (approximately 250 kW for a bus). Of course, the grocery store employee presumably could save enough money for a transatlantic flight, and leisure travel options admittedly have little bearing on issues of basic equity. Yet we cannot properly appreciate the widening social and economic divisions that characterize the contemporary world without grasping the implications of the ability to harness so much power for private use.

Even as the energy from oil and coal helped generate the affluence that made possible the broadening of the American middle class in the 1950s, it was also paving the way for a greater segregation of the poorer classes: Automobiles enabled the newly wealthy middle class to uproot to the outlying suburbs, leaving inner-city neighborhoods to deteriorate as tax-paying businesses moved out and property values fell.

This trend of suburbanization and income inequality has become even more exaggerated in recent years. Gated and walled communities are now ubiquitous in the United States.[9] Nor is it confined to the wealthier nations. In his book on corporate power, David Korten describes the rise of a new international globetrotting elite – a select club of bankers, bureaucrats, financiers, and businessmen who divide their time between the financial capitals of the world.[10] Recounting two visits to Pakistan as the guest of some of the country's most successful businessmen, Korten notes how detached they were from any concern with local, regional or national interests, and how little they seemed to know about their fellow citizens. In the same chapter he describes the view from his office window in a high-rise building in Makati, the financial center of the capital of the Philippines, Manila:

> Almost any time of the day I could see one or more private helicopters ferrying Manila's business elites to and from the tops of these high-rise buildings far above the cars stalled in Manila's legendary traffic jams and the lines of the carless commuters waiting amid thick diesel fumes for public transportation. On the other side of Manila, thousands of less fortunate Filipinos had built their shacks of scavenged materials on top of Smokey Mountain, a steaming garbage dump, and made their livings picking through the stinking mountain of garbage for bottles, bits of plastic, and other salable items.[11]

This image is emblematic of the social and economic divisions fostered by conventional energy systems. Business elites are insulated from the crowds by their helicopters, but social isolation of a different sort is the fate of the masses that cannot afford access to energy and find themselves without the means or opportunity to improve their economic situation.

It is widely if not universally accepted in the economic literature today that whatever the gains brought by globalization, it is has greatly widened the gap between rich and poor.[12] A case can be made that more than any other contributing circumstance, the extraordinary but unequally distributed growth in global energy consumption has been the driving force widening this gap. Globalization would not have been possible without the development of the energy-intensive technologies of modern communication and transportation. Yet modern vehicles and computers, and the energy systems that power them, are only available to those who can afford them. As Ivan Illich observed in the context of the energy crisis of the 1970s, such technologies "create distances for all, and shrink them for only a few. A new dirt road through the wilderness brings the city within view, but not within reach, of most Brazilian subsistence farmers. The new expressway expands Chicago, but it sucks those who are well-wheeled away from a downtown that decays into a ghetto."[13] Only by factoring in the wide discrepancies in energy use between people and countries can we gauge the inequalities in economic opportunity that characterize the world today.

Various studies demonstrate that energy inequality is a fact of life for both rich and poor countries alike. One study appraised the equity of energy use in El Salvador, Kenya, Norway, Thailand, and the United States.[14] Even the most equitable country, Norway, saw half of residential electricity being used by only 38 percent of customers; in the United States half of household electricity was used by 25 percent of households; in El Salvador, 15 percent; Thailand, 13 percent; and Kenya, 6 percent. Other studies have confirmed similar trends at the national level in Greece,[15] Mexico,[16] and the United Kingdom,[17] as well as at the village and household level in India.[18] One study looked at large disparities in the United Kingdom in the use of energy-intensive technologies like clothes washers, clothes driers, refrigerators, and freezers.[19] Another study looked at fossil fuel consumption in the United Kingdom from 1968 to 2000; not surprisingly, it found that higher incomes result in much greater consumption of fuel, car use, recreation, and international travel.[20] Yet another UK study found a clear relationship between income and car ownership, with the richest decile 11 times more likely to have the use of a private car than households in the poorest decile, where less than 1 in 10 own their own automobile.[21] In Australia, middle- and upper-income households consumed as much as four times the amount of fuel, light, power, and transport services than lower-income homes.[22]

For perhaps obvious reasons, energy inequality is most pronounced in the developing world. One study of energy use in Latin America and the

Caribbean found that the highest quintiles of households consumed 3 to 21 times more energy than the lowest quintiles.[23] Again, it is necessary to point out that energy use does more than reflect already existing social inequalities – it also tends to increase the distance between rich and poor. The study found, for example, that in all countries analyzed, the poor consumed less energy than other strata of households, but spent more of their available income on it. The price per unit of energy was usually higher, since poor households have difficulty accessing electricity grids or liquid fuels, which have higher energy densities. Another comprehensive analysis looked at middle-income consumption patterns in 14 countries and found extreme asymmetries in wealth. In India, the upper middle-income households account for less than one-eighth of the population, but account for 85 percent of private spending. The study noted that the per capita energy consumption in these more affluent homes is 15 times greater than that of the rest of India's population.[24]

Rural households tend to be poorer and consume less energy than urban households, and in rural areas fuelwood, the most common source of energy, is usually harvested in unsustainable ways, with severe impacts on forest health and the health of those using it, something we discuss below. However, the problem of energy equity and access is not confined to rural areas. One recent investigation of 34 cities in Bolivia, Botswana, Burkina Faso, Cape Verde, Haiti, India, Indonesia, Mauritania, Philippines, Thailand, Yemen, Zambia, and Zimbabwe concluded that, "the poor in urban areas of developing countries face special problems in meeting their basic energy needs."[25] The study noted that most urban poor are migrants and continue to rely on traditional fuels they collect on the periphery of urban areas, and also that they pay higher prices for usable energy because of the inefficiency of stoves and lamps.

While it should be clear from a review of these studies that energy inequality affects both rich and poor nations, and shares many characteristics in both contexts, the academic literature on this subject distinguishes between fuel poverty (characteristic of wealthy countries) and energy poverty (affecting developing countries). We now turn to a discussion of these two types of energy inequality.

Fuel poverty

Fuel poverty is a term that refers to households that have to spend over 10 percent of their income on energy services necessary to meet basic needs.[26] The term originated in the United Kingdom – unsurprisingly, given the large number of damp, poorly insulated, and drafty houses[27] – and fuel poverty has been described as a "peculiarly British public health scandal and an affront to human rights."[28] This framing is slightly misleading, however, as fuel poverty is found throughout the world. In recent years the issue has received increased attention in Eastern Europe, the Asia Pacific, Africa and even North America.[29] Fuel poverty today is escalating everywhere, due to the coincidence of rising fuel prices with decreasing household purchasing power; it is probably more helpful to think of it as a global problem, rather than limiting the term only to wealthy countries.

Fuel poverty typically results in inadequately heated houses (though this is not the only consequence), with a wide range of associated health impacts, including increased risk of respiratory and circulatory disease in adults, asthma in children, thousands of excess winter deaths among the elderly, and increased risk of mental health illness, and social isolation.[30] The groups most vulnerable to fuel poverty are low-income households with children, the elderly, the disabled, and the long-term sick.[31] According to statistics from the Department of Energy and Climate Change (DECC) in the United Kingdom, from 2004 to 2009 the number of English households struggling with fuel poverty doubled from 2 million to 4 million.[32] During this period, domestic electricity prices increased by over 75 percent and gas prices by over 122 percent.[33] Stagnant incomes have dramatically worsened the situation. A report published in 2012 noted that over 6 million households could not keep their homes warm.[34] Fuel poverty is an even greater problem in Southern and Eastern European states: over 30 percent of households in Bulgaria, Cyprus, and Portugal do not have enough money to heat the home adequately.[35]

Rising prices for fuel have a disproportionate impact on low-income families, who must pay a larger share of their income to meet their energy needs. One UK study in 1993 found that households with incomes in the top 20 percent spent only 4.2 percent of their budget on fuel, while households in the bottom 20 percent spent 12.1 percent of available income.[36] The burden for low-income households was nearly three times greater than that for wealthy households. The inequity of this situation is compounded by the fact that the poor do not have the resources to make cost-saving capital investments on improving household energy efficiency. As another UK study stated, "the prospect for low-income groups is of continuing dependence on increasingly expensive electricity and gas supplies, making them an 'energy underclass' at continued or increased risk of fuel poverty."[37] According to a report prepared for Consumer Focus, rising energy bills in the United Kingdom will most likely result in the number of households in fuel poverty exceeding 9 million by 2016.[38]

Most severely, fuel poverty leads to "excess winter mortalities," quite literally killing people who go without essential heat. One epidemiological study looked at 11 industrialized countries in both the Northern and Southern Hemispheres and found a clear correlation between the winter months and unusually high rates of mortality, a trend they plotted in Figure 4.2.[39] Table 4.1 shows the average excess winter deaths – defined as the extra deaths in the four winter months in comparison with the previous and succeeding four months – across a dozen countries and calculates that these amount to 278,409,[40] exceeding the global number of deaths (about 166,000) attributed by the World Health Organization (WHO) to climate change.[41] To be fair, the nature of these deaths is different: excess winter deaths occur primarily among the elderly, whereas the deaths from climate change affect all ages. But

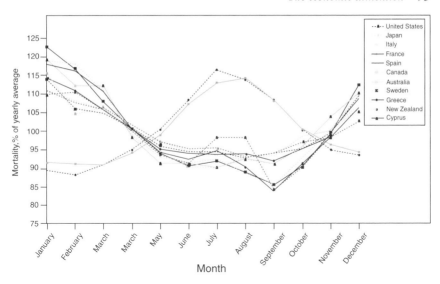

Figure 4.2 Monthly mortality for various developed countries as a percent of yearly average, 2009

Source: World Health Organization.

by merely counting the numbers, fuel poverty is in one way as egregious to society's overall health as climate change.

Returning to a theme we raised earlier in the chapter, social policy advocates concerned with poverty issues often argue that any kind of energy or carbon tax is regressive because it negatively affects the poorest in society.[42] When the

Table 4.1 Average excess winter mortality in 12 countries, 2008

Country	Population (millions)	Average excess winter mortality
Australia	19.5	6,973
Canada	30.3	8,113
Cyprus	0.8	317
France	59.5	24,938
Greece	9.5	5,820
Italy	54.4	37,498
Japan	127.9	50,887
New Zealand	3.5	1,600
Spain	37.5	23,645
Sweden	8.8	4,034
United Kingdom	60.9	36,700
United States	287.3	77,884
Total		278,409

Source: Falagas et al. and Day and Hitchings.

United Kingdom announced a new carbon tax in the spring of 2013, the Institute for Public Policy Research issued a press release claiming that the new tax would increase the price of wholesale electricity by around 17 percent in 2015 and would push another 30,000 to 60,000 households into fuel poverty.[43] One of the justifications for hydraulic fracturing in the United States is the downward pressure the new supply of natural gas is exerting on gas and electricity prices, which alleviates the burden on millions of people in the lower income brackets.[44] Similar equity considerations are brought to bear on the question of whether government policies should attempt to keep the price of gas at the pump artificially low, especially in North America where suburban sprawl and the lack of public transport alternatives mean that many people are entirely dependent on their cars. As we have already discussed, we cannot lightly dismiss these arguments if we are concerned with energy justice.

There is no question that some environmental and energy taxes have made the poor worse off. As a comprehensive study of this question concluded, "Unless environmental taxes are specifically designed to be progressive, they probably won't be."[45] But that is precisely the point: the regressive or progressive impact of a tax depends entirely on how it is designed. If the revenue from carbon taxes were invested in measures such as insulation and efficient boilers for households struggling to pay their bills, it could reduce fuel poverty in the United Kingdom by 90 percent, while generating up to 71,000 jobs by 2015 and boosting the economy throughout the country.[46] A study looking at the relationship between decentralized energy systems and fuel poverty found that a combination of energy efficiency measures and distributed microgeneration (with technologies such as ground source heat pumps and solar panels) could improve access to affordable energy for fuel poor households, but only if central and local governments adopted programs that addressed the insurmountable obstacle for low-income families of the relatively high capital costs of such improvements.[47] In the end, such measures will be necessary for the equitable distribution of energy services. Even if shale gas temporarily decreases the price of electricity in the United States, prices for oil and gas are primarily driven by the international market, and global demand will continue to grow, inexorably driving up the price of energy.

Energy poverty and drudgery

The degree of energy inequality in most developing countries dwarfs that found in the developed world. If we visualize global energy consumption as a pyramid composed of around 7 billion people, the top 500 million consume an average per year of 10,000 kWh/household and the bottom 1.3 billion live with no electricity at all. Of the remaining 5 billion people, the top half consumes an annual average of 5,000 kWh/household and the bottom half an annual average of 1,000 kWh/household (about the same as what the average American home consumes in a month). Approximately 2.7 billion people are entirely dependent on wood, charcoal, and dung for

Table 4.2 Number and share of population without access to modern energy services, 2009

		Without access to electricity		Dependence on traditional solid fuels for cooking	
		Population (million)	*Share of population (%)*	*Population (million)*	*Share of population (%)*
Africa		**587**	**58**	**657**	**65**
	Nigeria	76	49	104	67
	Ethiopia	69	83	77	93
	Congo	59	89	62	94
	Tanzania	38	86	41	94
	Kenya	33	84	33	83
	Other sub-Saharan Africa	310	68	335	74
	North Africa	2	1	4	3
Asia		**675**	**19**	**1,921**	**54**
	India	289	25	836	72
	Bangladesh	96	59	143	88
	Indonesia	82	36	124	54
	Pakistan	64	38	122	72
	Myanmar	44	87	48	95
	Rest of developing Asia	102	6	648	36
Latin America		**31**	**7**	**85**	**19**
Middle East		**21**	**11**	**0**	**0**
Developing Countries		**1,314**	**25**	**2,662**	**51**
World		**1,417**	**19**	**2,662**	**39**

Source: International Energy Agency.

domestic cooking (see Table 4.2).[48] Two-thirds of the people without any access to electricity live in 10 countries – four in Asia and six in sub-Saharan Africa – and half of the people lacking clean cooking facilities live in only three countries – China, India, and Bangladesh.[49]

We have already briefly touched upon some of the consequences of inadequate access to modern energy services in developing countries. Energy poverty (the term commonly used to refer to a lack of access to electricity and dependence on solid biomass fuels for cooking and heating) dramatically reduces opportunities for developing the capabilities that lead to a flourishing human life in a globalized world; in more extreme instances, it makes it difficult to meet survival needs. Effective and financially rewarding participation in the global economy is restricted without access to energy; moreover, political representation for those deprived of modern communication technologies is likewise limited.[50] Energy is power, and in a globalized world, human agency is diminished as a consequence of energy poverty.

The problems associated with fuel poverty in the developed world are the same but more exaggerated in the developing world. Poor households tend to spend 20 to 30 percent of their annual income on energy fuels, and the indirect costs associated with collecting and using that energy – such as loss of time, injury, and healthcare expenses – can account for an additional 20 to 40 percent. On average, the poor pay eight times more for the same unit of energy than higher income groups.[51] In the worst cases, the poorest households are compelled to spend 80 percent of their income acquiring cooking fuels.[52]

The socioeconomic implications of energy poverty extend beyond income, however. The health concerns related to energy poverty include indoor air pollution, physical injury during fuel wood collection, and the consequences of a lack of refrigeration and health care services. The first of these is the most serious by far. As can be seen in Table 4.2, just under half of the entire global population is dependent on wood, charcoal, and dung for cooking. In rural areas, 75 percent of households use solid biomass fuels for cooking.[53] The majority of households combust this fuel directly inside the home, typically without adequate ventilation. According to the World Health Organization:

> The inefficient burning of solid fuels on an open fire or traditional stove indoors creates a dangerous cocktail of not only hundreds of pollutants, primarily carbon monoxide and small particles, but also nitrogen oxides, benzene, butadiene, formaldehyde, polyaromatic hydrocarbons and many other health-damaging chemicals.[54]

The ill effects of burning solid fuels indoors are compounded by a number of factors. First, most homes are small and cramped and therefore concentrate the resulting fumes, leading to exposure levels to pollutants that can exceed 60 times the acceptable outdoor rate in North American and European cities.[55] Secondly, for obvious reasons, fuels are burned when people are present, whether cooking, eating, or sleeping. As women usually spend three to 7 hours a day in the kitchen, often accompanied by small children, indoor air pollution has a disproportionate impact on women and children.[56] The health impacts of indoor air pollution are devastating, and include acute respiratory infections, tuberculosis, chronic respiratory diseases, cardiovascular disease, lung cancer, asthma, low birth weights, diseases of the eye, and adverse pregnancy outcomes.[57]

Indoor air pollution ranks *fourth* on the global burden of disease risk factors at almost 5 percent, coming after only high blood pressure (almost 8 percent), tobacco smoking and second-hand smoke (about seven percent), and alcohol use (about 6 percent).[58] This places it well ahead of physical inactivity, obesity, drug use, and unsafe sex, as Figure 4.3 depicts. In India

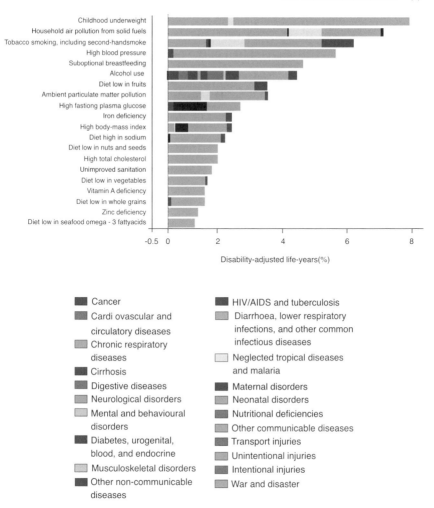

Figure 4.3 Burden of disease attributable to 20 leading risk factors (percent of disability adjusted life years globally, for both sexes), 2012

Source: Lim et al., 2012.

and all of South Asia, cook stove smoke is the *highest* health risk factor, ranking above smoking and high blood pressure. It is second for sub-Saharan Africa, third for Southeast Asia, and fifth in East Asia. Air pollution from conventional cook stoves is therefore responsible for 4 million deaths each year, 3.5 million direct premature annual deaths and 500,000 from "second-hand cook fire smoke."[59] In aggregate, such air pollution amounts to almost 11,000 deaths per day, or almost eight deaths per minute. The cost of this burden to national healthcare systems, not reflected in the price of fuel wood or energy, is $212 billion to $1.1 trillion.[60]

Not only are women and children most vulnerable to the pollution associated with energy poverty, they also suffer disproportionately from the hazards of fuel collection. These range in severity from cuts and back injuries to sexual assault and murder. Owing to increasingly scarce resources, women can be forced to spend hours each day gathering fuel; because they often have no choice but to take their children with them, the children are exposed to the same dangers.[61] A typical woman in India spends 40 hours per month collecting fuel; many women have to walk a round trip of more than 6 kilometers 15 times per month, carrying heavy loads of firewood on the return journey.[62] In aggregate, this means that women in India spend 30 billion hours per year (82 million hours per day) gathering firewood.[63] In Addis Ababa, Ethiopia, 10,000 fuel wood carriers supply one-third of the wood needed by the city for cooking and heating. These women suffer falls, bone fractures, headaches, anemia, internal body disorders, and miscarriages from carrying loads often equal to their body weights. In addition to the variety of injuries that are sustained in actually carrying fuel, fuel collection also exposes women and children to physical and sexual violence. In Somalia, there are hundreds of documented cases of women being raped while gathering fuel.[64]

More widely distributed access to affordable electricity is one of the obvious solutions to the adverse health impacts of energy poverty. Not only would this do away with the necessity of gathering, carrying, and burning solid biomass fuels, it would also enable developing countries to improve dilapidated healthcare systems. But little has been done to improve the supply of electricity in much of the developing world. Over half of the poorest countries in sub-Saharan Africa face "crippling electricity shortages." With the exception of South Africa, the remaining 700 million people living within this vast area have access to approximately the same amount of energy as do the 38 million citizens of Poland. According to Lawrence Musaba, the manager of the Southern African Power Pool, there has been "no significant capital injection into generation and transmission, from either the private or public sectors, for 15, maybe 20 years." Maintenance of some transmission grids within Africa is so deplorable that energy suppliers are paid for only 60 percent of generated electricity. And generation is no better than transmission. In 2007, only 19 of the 79 power plants in Nigeria were actually operational, and blackouts cost the economy $1 billion per year.[65]

Reliable distribution of energy services is not the only issue, however. As we have been emphasizing throughout this chapter, affordable energy is equally important for energy justice. The current controversy in South Africa over the construction of the Medupi and Kusile coal-fired power plants illustrates this problem well. The state-owned electric utility Eskom awarded contracts to build these two power plants in 2007, following severe power shortages in South Africa. Both power plants will have a gross generating capacity of approximately 4,800 MW. Estimated costs for the Medupi power station were $17.8 billion, though the plant (the first to be constructed) has not yet been completed and costs continue to exceed projections.[66] The Medupi plant alone will emit approximately 30 million metric tons of carbon dioxide-equivalent greenhouse gases per year, more than the 63 lowest-emitting countries combined, and will be the seventh largest coal-fired power plant in the world.[67] Government apologists for the power plant have argued that it is necessary to support economic growth and development in South Africa and will help alleviate poverty and in particular fuel poverty.[68] Critics of the project point to the fact that energy-intensive industries like mining and aluminum smelting use a large proportion of South African electricity,[69] and suggest that the real purpose behind Medupi and Kusile is to sustain low-cost energy for multinational corporations like BHP Billiton, Anglo American Platinum, and ArcelorMittal.[70] The power plants, critics allege, will simply ensure a "further entrenching of an export-orientated economy in raw materials," an "economic model that has consistently failed, for the last hundred years, to eradicate poverty in the country."[71]

There is no doubt that South Africa needs to do something to address its power shortages. In early 2008 it experienced a severe power crisis, and economic growth continues to put a strain on the electricity grid, threatening the power supply to many industries.[72] However, 45 percent of the rural population (approximately 12.5 million people) has no access to electricity, and the Medupi plant will do nothing to alter this fact.[73] South Africa has abundant renewable energy resources, including solar and wind power, which have been assessed by the government, though no serious action has been taken to develop these more distributed sources of energy.[74] South Africa has also experienced fast-rising electricity costs in recent years: From 2005 to 2011, electricity prices rose by 109 percent.[75] In 2008, in response to a request for a 60 percent tariff increase from Eskom to help finance the Medupi plant, the National Energy Regulator of South Africa (NERSA) approved a tariff increase of 27.5 percent; more recently, NERSA approved an annual increase of 16 percent for the next three years.[76] Households now pay an average of R1.40 (about 14¢) per kWh,[77] a retail rate higher than that of the United States, and fuel poverty in South Africa has become an increasingly divisive issue.[78] Despite the outrage expressed by increasing numbers of people, and protests at both Medupi and Kusile, extractive industries in South Africa continue to pay below market prices for power. In March 2013, a court order ratified by the Supreme Court of Appeal compelled Eskom to divulge details of its deals with BHP Billiton, dating back to 1992. Contracts released by Eskom revealed that one of BHP Billiton's aluminum smelters, Hillside, currently pays less than one-fifth the price that average customers pay.[79]

There are thus good reasons to question whether further developing conventional energy systems will deliver a more equitable distribution of energy services, especially in countries where there are large rural populations remote from central power stations and distribution networks. An enormous sum of money has been invested in the Medupi power plant, a sizeable portion of it ($3.7 billion) coming from the World Bank, which has a mandate to develop clean energy.

These funds could equally be used to invest in demand-side management, the development of renewable energy sources, and the construction of smaller-scale generation plants distributed over a wider geographic area. If this approach were implemented properly, it could meet the rising energy needs of South Africa, distribute energy services more equitably, and in the process create thousands of "green" jobs for the currently unemployed (e.g., installing every house with a solar home system). The Medupi and Kusile power plants will ensure a reliable supply of base load power for mining and smelting interests in South Africa. To the extent that the economy there is dependent on these extractive industries, this will also serve the interests of South Africa. But there is a high cost to be paid for this reliability, and it is not being paid by the smelting companies.

Resource depletion and rising energy prices

The question of peak oil has been debated for decades now and we have no intention of weighing in on this controversial topic. One thing is clear, though: Reserves of conventional, easy to access oil and gas are dwindling. Coupled with the exponential growth in global demand for fossil fuels, this means that the era of cheap oil is finished. According to a world energy outlook recently released by British Petroleum, the long run price movements for fossil fuels show a steady upwards trend. Average annual real oil prices between 2007 and 2011 were 220 percent above the average for the period 1997–2001. The increase for coal prices was 141 percent, and for gas prices, 95 percent.[80]

High energy prices and technological innovation have unlocked previously inaccessible reserves of unconventional resources: shale gas, tight oil, shale oil, deepwater oil, and gas and heavy crude from oil sands. Moreover, technological advances are making it possible to extract a greater percentage of oil from proven reserves, further increasing global supply. But while the "shale revolution" has implications for the debate over peak oil, it does not alter the fact of rising fossil fuel prices, for the simple reason that unconventional fossil fuel resources are only accessible because prices are high. If prices were to drop precipitously, these resources would no longer be economical to extract. For this reason, British Petroleum suggests that most likely the world will see a peak in demand before it reaches a peak in supply. Already oil is following what British Petroleum calls a "long run decline in its market share," with the demand for oil increasingly concentrated in the transportation sector, where it is most highly valued.[81]

The implications of these trends for energy justice are clear. As the prices for fossil fuels rise, the inequities of the conventional global energy system will only intensify: Access to energy services will become increasingly restricted to those who can afford it. Moreover, as the global economy remains dependent on fossil fuels, and is set to continue so for the foreseeable future, rising prices will have disastrous knock-on effects for the most vulnerable people throughout the world, driving up food prices and pushing greater numbers of people into destitution.

Fossil fuels: A finite resource

The world's four primary energy fuels are concentrated in a dramatically low number of countries, creating significant patterns of import dependence, and these conventional reserves are declining. The world's 1.5 trillion barrels of proven oil reserves[82] are for the most part located in volatile regions of the world.[83] The geographic distribution of oil reserves is staggering, as Saudi Arabia (262.5 billion barrels) and Venezuela (211.2 billion barrels) control more than 30 percent of all the world's crude, whereas 80 countries have reserves of 10 billion barrels or fewer and 117 countries have virtually no oil reserves.[84] The three largest oil and gas companies in terms of output – Saudi Arabian Oil Company, Gazprom, and National Iranian Oil Company – produced 28.6 billion barrels of oil equivalent a day in 2012.[85] From 2002 to 2011, the oil industry's annual spending on exploration and production has increased by a factor of four, yet oil production is up as a result by only 12 percent.[86]

The geopolitical distribution of other conventional energy resources, such as natural gas, coal, and uranium, is equally concentrated. Figure 4.4 provides a graph of conventional energy reserves by country.[87] Eighty percent of the world's oil can be found in nine countries that have only 5 percent of the world population and 5 percent of gross domestic product (GDP); 80 percent of the world's natural gas is located in 13 countries with 12 percent of the population and 26 percent of GDP; and 80 percent

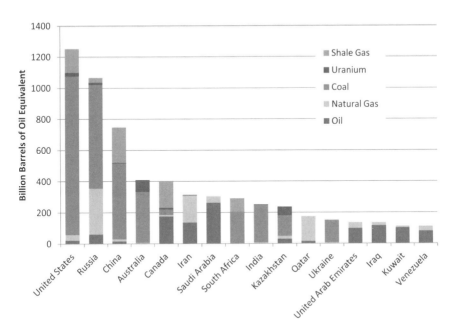

Figure 4.4 Conventional energy reserves for leading countries, 2012

Source: Sovacool and Valentine.

Table 4.3 Life expectancy of conventional fossil fuel and uranium resources, 2012

	Proven reserves	*Current production*	*Life expectancy (years)*		
			0% annual production growth rate	*1.6% production growth rate*	*2.5% production growth rate*
Coal	930,400 million short tons	6,807 million short tons	137	85	61
Natural Gas	6,189 trillion cubic feet	104.0 trillion cubic feet	60	42	37
Petroleum	1317 billion barrels	30.560 billion barrels	43	33	30
Uranium	4,743,000 tons (at $130/kgU)	40,260 tons	118	67	56

Source: Sovacool and Valentine.

of the world's coal is located in 6 countries (though these countries have 45 percent of the population and 46 percent of GDP). Some of the same countries are among the 6 that control more than 80 percent of global uranium resources.

Though reserves of conventional fossil fuels may seem vast, growing world-wide demand for electricity and mobility threatens to exhaust them in the foreseeable future. The global economy is transitioning from a position of abundant fossil energy supplies to a largely resource-constrained future. World energy demand is expected to expand by 45 percent between now and 2030, and by more than 300 percent by the end of the century.[88] Table 4.3 summarizes three production scenarios for conventional fossil fuels. In the first scenario, current production levels remain constant. In other words, expansion of production capacity falls further behind the pace of global demand increases, suggesting massive escalation of resource prices. In the second scenario, production capacity expands by the same rate (1.6%) as short-term projections for demand increases. In the third scenario, production capacity expands by the same rate (2.5%) as long-term projections for demand increases. As the table demonstrates, if long-term projections remain valid, oil reserves will be exhausted in 30 years, natural gas in 37 years, and coal in 61 years.

To date, the world has consumed about 1 trillion barrels of oil. Although oil production has been ongoing since the nineteenth century, 99.5 percent of all production has occurred within the last 60 years. Today we consume 30 *billion* barrels of oil per year, and future demand for oil is projected to grow at more than twice the historic rate since 1980. It is estimated that there are remaining conventional reserves of approximately 1 trillion barrels, which is enough to last around three decades at today's consumption rate.

As already stated, high prices of fossil fuels driving advances in technology have altered this picture of declining reserves to a certain extent. According to current estimates, there are technically recoverable global resources of 240 billion barrels of tight oil and 200 trillion cubic meters of shale gas.[89] Yet as British Petroleum states in its recent energy outlook, whether these "technically recoverable" resources can be extracted depends on "above ground" factors, such as available technology, financial resources, and a permissive regulatory framework (meaning one that continues to ignore the perils of climate change).[90] In all likelihood, these factors will mean that a considerable percentage of the supply of unconventional resources remain buried in the ground. Moreover, in terms of the economic dimension of energy justice, the production of shale gas, tight oil, oil sands, and Arctic offshore oil and gas does not change the fundamental reality of a rising trend in fossil fuel energy prices.

It thus remains unclear how long the world can produce enough *affordable* oil (and coal, natural gas, and uranium) to meet growing global demand. What is certain is that a growing percentage of global oil demand will be met using resources from the Middle East, since almost half (45 percent) of the world's proven reserves of conventional oil are located in Saudi Arabia, Iraq, and Iran. If transportation remains oil dependent and if surplus oil production in the world is limited to one or 2 million barrels per day,[91] oil-importing nations across the world will continue to be vulnerable to oil price shocks and volatility, as has been the case since the Yom Kippur War and oil embargo of 1973. The likely geographic pattern of oil production and consumption over the next two decades suggests that oil dependence in Europe, China, India, and other Asian countries could grow rapidly, each country relying on imports to meet more than 75 percent of oil demand by 2030.[92]

Rising energy prices and volatile markets

Resource concentration makes the prices of fossil fuels volatile. As Figure 4.5 shows, oil prices have been incredibly volatile over the past few decades. Figure 4.6 shows that the price of natural gas at the Henry Hub in Louisiana has been similarly volatile, jumping from $6.20 per million BTUs (MMBtu) in 1998 to $14.50 per MMBtu in 2001, then dropping precipitously for almost a year only to rebound yet again, and collapse, again, with the accelerated production of shale gas.[93] In August and September of 2005, two single storms in the Gulf of Mexico – Hurricanes Katrina and Rita – brought down 29 percent of refinery capacity in the United States, shut off 66 percent of oil and 54 percent of natural gas production in the Gulf of Mexico, and disrupted 16 natural gas processing plants, with calamitous results for oil and gas prices.[94]

The global repercussions of rising oil prices have been severe. One study noted that the rising oil prices of 2010 and 2011 placed an additional 42 million people in the Asia-Pacific region into poverty.[95] A second study assessed the close connection between rising oil prices and food prices, and documented an almost perfect relationship between the two. It is important

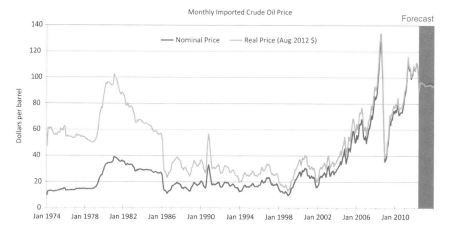

Figure 4.5 Global crude oil prices, 1987 to 2012

Source: U.S. Energy Information Administration.

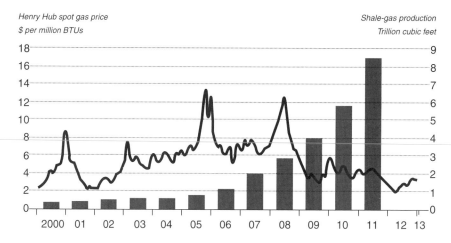

Figure 4.6 Henry Hub spot prices for natural gas, 2000 to 2013

Source: U.S. Energy Information Administration.

to pause here and emphasize a point we have already touched upon a few times in this chapter. Rising oil prices do not affect only those who consume oil. Because the global economy remains dependent on oil, rising prices have an indirect effect on the prices of basic commodities, and thereby affect the livelihoods even of people who have no access to energy. Higher oil prices result in rising input costs for agriculture, such as oil-based fertilizers and

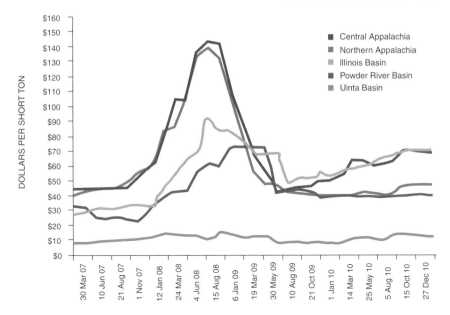

Figure 4.7 Coal commodity spot prices, 2007 to 2010

Source: Freese et al.

fuel for motorized and mechanized equipment, as well as a greater demand for biofuels, which then divert agricultural feedstocks to produce fuel rather than food. Both factors create higher food prices, and between 2004 and 2007 were responsible for increasing the number of malnourished people in the world from 848 million to 923 million.[96] We have to question the adequacy of the ethics of the marketplace when the price mechanism for oil creates food shortages for millions of people who might never even have driven a car.

Figure 4.7 shows that coal prices have also been volatile, jumping from $50 per short ton in Central Appalachia in November 2007 to $140 per short ton in August 2008, almost tripling in nine months.[97] Transportation bottlenecks and demand surges in developing countries such as India and China were partly to blame. From 2001 to 2006, coal use around the world grew by 30 percent, and 88 percent of this increase came from Asia (with 72 percent of the increase from China alone).[98] In 2004, China shifted from being a net exporter to a net importer of coal, creating a severe shortage of ocean-worthy bulk carriers and causing the global price of coal (as well as other commodities such as iron, ore, and steel) to increase dramatically as freight prices soared.[99] Other contributing factors of volatile coal prices include: (a) dwindling reserves of coal in some parts of the United States and Europe; (b) constricted rail service; (c) flooding and hurricanes hitting barge routes; (d) bankruptcies and resulting consolidation and restructuring within the industry; (e) permitting, bonding, and insurance issues; (f) mine closures; (g) more stringent environmental regulations concerning mine planning and the posting of reclamation bonds; and (h) restrictions on mountain-top removal.[100] A comprehensive International Energy Agency study in 2003 revealed that more than half of coal-powered plants reported significant price fluctuations in coal, varying by almost a factor of 20.[101]

Even the predominant fuel for nuclear power plants, uranium, has exhibited considerable volatility. Uranium fuel costs have shown recent escalation and volatility, which negatively impact nuclear project economics, making them riskier than many existing alternatives. For example, Figure 4.8 depicts uranium spot prices from 1965 to 2009 and reveals two astronomical spikes: one in the 1970s and 1980s as a large number of commercial nuclear power plants began operation, and a second one post-2004 caused by an expected renaissance in the nuclear industry and resulting constraints in supply.[102] The cost of uranium, for instance, jumped from $7.25 per pound in 2001 to $47.25 per pound in 2006, an increase of more than 600 percent. The Nuclear Energy Agency reports 200 metric tons of uranium are required annually for every 1,000 MW reactor and that uranium fuel accounts for 15 percent of the lifetime costs of a nuclear plant, meaning that price spikes and volatility can cost millions of dollars, affecting the profitability of a plant.[103] Uranium price volatility has been influenced heavily by the unexpected introduction of secondary supplies

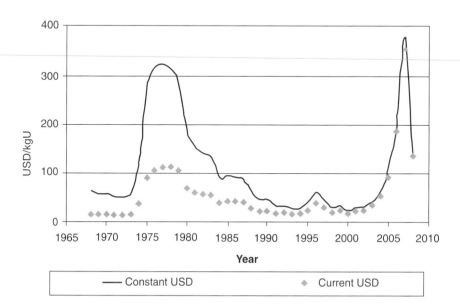

Figure 4.8 Uranium market prices, 1965 to 2009

Source: International Atomic Energy Agency.

and gluts in the market, connected in part to sudden increases in supply from cancelled and shut down reactors and the dilution of highly enriched uranium from surplus nuclear weapons.

Conclusion

Economic opportunities in the modern globalized world are extremely limited without access to energy. In recent decades, the most remarkable increase in energy use has been in the developing world; industrialization and economic growth – fueled by the consumption of coal, oil and gas – have raised millions of people out of poverty and destitution. However, even as global energy consumption has increased exponentially, an equitable distribution of reliable and affordable energy services remains as elusive as ever. Billions of people live with inadequate or non-existent access to energy services, and are thereby excluded from meaningful participation in the global economy and the opportunity to live a flourishing human life.

Energy inequality is not only characteristic of the developing world; it is a fact of life even in the wealthiest countries. This latter circumstance more than any other should make us question the prevailing assumption that the way forward is simply more of the same – more coal power plants, more oil and gas production, more centralized grids. Wealthy nations have benefited from reasonably reliable national grids, relatively cheap supplies of oil and gas, and adequate generating capacity for more than half a century, and yet fuel poverty is increasingly narrowing opportunities and lowering quality of life for millions of ordinary people within these countries.

More of the same cannot rectify this situation, because a significant part of the problem is that prices for fossil fuels, which remain the dominant source of energy throughout the world, are ineluctably rising, driven by rapidly increasing global demand and diminishing reserves of easily accessible oil and gas. Even the "shale revolution" – along with all the other unconventional fossil fuel resources that have become accessible through new innovations in technology – will not alter this situation to any significant degree, because the costs of extraction are such that they only make economical sense if prices are high. As long as we choose to continue powering our economic activity with fossil fuels – which are rivalrous goods, as opposed to the non-rivalrous and abundant sources of renewable energy – the distribution of *affordable* energy services will only become more inequitable with time, regardless of how many new shale fields we exploit and coal plants we construct.

For the reasons just outlined (and summarized below in bullet points), we believe that the affirmative principle of energy justice requires a fundamentally rethinking of the dominant model of global energy policy.

- The ongoing construction of conventional energy systems in developing countries tends to ignore the issue of how energy services are distributed;

while this growth in energy consumption helps sustain GDP growth in any given country, it does not always increase the overall wealth of the country, and it does not benefit a large portion of the population. GDP is not the only measure of a nation's economic well-being.

- More energy is not necessarily the solution to energy inequality when the type of energy provided puts it out of the price range of those who are most in need of it; it is not only a question of more, but what kind of energy, how it is distributed, and to whom. Steadily rising energy prices make this problem more acute.

- Fuel poverty, a term that refers to households that have to spend over 10 percent of their income on energy services necessary to meet basic needs, is a growing problem throughout the world; the costs associated with fuel poverty can be severe, and include increased risk of respiratory and circulatory disease, asthma, thousands of excess winter deaths and increased risk of mental health illness, and social isolation.

- Fuel poverty affects the poorest and most vulnerable groups within society, in both developed and developing countries; increased energy production usually fails to alleviate the situation of women, children, the elderly, the infirm, or minorities, as their plight is determined not by a shortage of supply (as can be seen, for instance, in the United States), but by rising energy prices.

- The degree of energy inequality within and across countries today is extreme: 1.3 billion people have no access to electricity, and 2.7 billion rely on biomass to meet their cooking and heating needs.

- Energy poverty typically results in fewer educational opportunities, less access to fertile land and other natural resources, poorer health, negligible political representation, limited economic opportunities and inadequate access to health services.

- The health consequences of energy poverty are severe: indoor air pollution (IAP) is fourth on the global burden of disease risk factors; globally, approximately 4 million people die each year due to air pollution from conventional cook stoves; IAP (like many energy burdens) affects women and children disproportionately.

- Global demand for fossil fuels is increasing exponentially, and conventional reserves are becoming exhausted, driving up prices.

- The world's primary conventional energy fuels are concentrated in a relatively small number of countries, which creates significant patterns of import dependence; combined with diminishing reserves, and growing concerns over energy security, the potential for geopolitical instability is increasing, as is the possibility of resource conflicts.

- The "shale revolution" has unlocked significant reserves of unconventional fossil fuels; however, as extraction of these reserves is capital-intensive, and therefore dependent on high energy prices to be cost effective, these newly accessible resources will not alter the fundamental picture of rising energy prices.

- Because the global economy is dependent on the energy sector, volatility in energy prices affects the entire economy, leading to spikes in commodity prices and other ripple effects that threaten to destabilize international markets.

Notes

1. Nicholas Georgescu-Roegen, "Energy and Economic Myths," in *Valuing the Earth: Economics, Ecology, Ethics,* ed. Herman E. Daly and Kenneth N. Townsend (Cambridge, MA: The MIT Press, 1993).
2. International Energy Agency, United Nations Development Programme, United Nations Industrial Development Organization, *Energy Poverty: How to Make Modern Energy Access Universal?* (Paris: OECD, 2010).
3. For an industry perspective on this issue, see British Petroleum's 2013 *Energy Outlook 2030,* http://www.bp.com/extendedsectiongenericarticle.do?categoryId=9048887&contentId=7082549. For a broader perspective, see also Jonathan Porritt, *Capitalism as if the World Matters* (London: Earthscan, 2005); Richard Heinberg, *The Party's Over: Oil, War and the Fate of Industrial Societies* (Gabriola Island, BC: New Society Publishers, 2003); and Colin J. Campbell and Jean H. Laherrere, "The End of Cheap Oil?" *Scientific American* (March 1998).
4. World Bank, "Data," http://web.worldbank.org/WBSITE/EXTERNAL/EXTOED/EXTRURELECT/ (accessed May 30, 2013).
5. See, e.g., Amory Lovins, *Soft Energy Paths: Towards a Durable Peace* (San Francisco: Friends of the Earth International, 1977).
6. International Energy Agency, *World Energy Outlook 2012,* http://www.worldenergyoutlook.org/ publications/weo-2012/.
7. Larry Copeland, "Driving Costs are Up 1.9%, AAA Finds," *USA Today,* April 30, 2012, 3A.
8. World Bank, "Data," http://data.worldbank.org/indicator/EG.USE.PCAP.KG.OE/countries/1W?display=default.
9. See Edward J. Blakely and Mary Gail Snyder, "Divided We Fall: Gated and Walled Communities in the United States," in *Architecture of Fear,* ed. Nan Ellin (New York: Princeton Architectural Press, 1997).
10. David Korten, *When Corporations Rule the World,* 2nd ed. (San Francisco: Berrett-Koehler Publishers, 2001), 107–120.
11. Ibid., 116.
12. The literature on this subject is vast. One of the more popular authors who deal with the topic is the Nobel prize–winning economist Joseph E. Stiglitz. See, for example, his *Globalization and its Discontents* (New York: W.W. Norton & Company, 2003) and *Making Globalization Work* (New York: W.W. Norton & Company, 2007).
13. Ivan Illich, *Energy and Equity* (London: Marion Boyars, 2009), 42–43.
14. Arne Jacobson and Daniel Kammen, "Letting the (Energy) Gini Out of the Bottle: Lorenz Curves of Cumulative Electricity Consumption and Gini Coefficients as Metrics of Energy Distribution and Equity," *Energy Policy* 33, no. 14: 1825–1832.

15. T. Rapanos Vassilis and L. Michael Polemis, "The Structure of Residential Energy Demand in Greece," *Energy Policy* 34 (2006): 31–37.

16. Jorge Alberto Rosas-Flores, David Morillon Galvez, and Jose Luis Fernandez Zayas, "Inequality in the Distribution of Expense Allocated to the Main Energy Fuels for Mexican Households: 1968–2006," *Energy Economics* (2010).

17. L. C. Hunt, G. Judge, and Y. Ninomiya, "Underlying Trends and Seasonality in UK Energy Demand: A Sectoral Analysis," *Energy Economics* 25 (2003): 93–118.

18. E. Fernandez, R. P. Saini, and V. Devadas, "Relative Inequality in Energy Resource Consumption: A Case of Kanvashram Village, Pauri Garwal District, Uttranchall (India)," *Renewable Energy* 30 (2005): 763–772.

19. A. Druckman and T. Jackson, "Measuring Resource Inequalities: The Concepts and Methodology for an Area-Based Gini Coefficient," *Ecological Economics* 65 (2008): 242–252.

20. E. Papathanasopoulou and T. Jackson, "Measuring Fossil Resource Inequality: A Longitudinal Case Study for the UK: 1968–2000," *Ecological Economics* (2010).

21. Stephen Tindale and Chris Hewett, "Must the Poor Pay More? Sustainable Development, Social Justice, and Environmental Taxation," in *Fairness and Futurity: Essays on Environmental Sustainability and Social Justice,* ed. Andrew Dobson (Oxford: Oxford University Press, 1999), 233–248.

22. Tran Van Hoa, "Quality of Consumption: Some Australian Evidence," *Economics Letters* 19 (1985): 189–192.

23. United Nations Development Program, *Contribution of Energy Services to the Millennium Development Goals and to Poverty Alleviation in Latin America and the Caribbean* (Santiago, Chile: United Nations, 2009).

24. Norman Myers and Jennifer Kent, "New Consumers: The Influence of Affluence on the Environment," *Proceedings of the National Academies of Science* 100, no. 8 (2003): 4963–4968.

25. Douglas F. Barnes, Kerry Krutilla, and William Hyde, *The Urban Household Energy Transition: Energy, Poverty, and the Environment in the Developing World* (Washington, DC: Resources for the Future, 2004).

26. Mary O'Brien, "Fuel Poverty in England," *The Lancet,* December 5, 2011.

27. Tindale and Hewett, "Must the Poor Pay More?" 238.

28. V. Press, "Fuel Poverty and Health: A Guide for Primary Care Organizations, and Public Health and Primary Care Professionals," National Heart Forum, http://www.heartforum.org/downloads/FPbook.pdf, 5.

29. See EU Fuel Poverty Network, "Working to Raise Awareness of Fuel Poverty," May 2013, http://www.fuelpoverty.eu.

30. National Audit Office, "Warm Front: Helping to Combat Fuel Poverty," Report by the Comptroller and Auditor General, HC 769 Session 2002–2003, June 25, 2003, 1. See also O'Brien, "Fuel Poverty in England"; J. Rudge and R. Gilchrist, "Excess Winter Morbidity among Older People at Risk of Cold Homes: A Population-Based Study in a London Borough," *Journal of Public Health* 27, no. 4 (2005): 353–358.

31. National Audit Office, "Warm Front," 1.

32. Department of Energy and Climate Change, *Annual Report on Fuel Poverty Statistics 2011,* 2011, http://www.decc.gov.uk/assets/decc/Statistics/fuelpoverty/2181-annual-report-fuel-poverty-stats-2011.pdf.

33. Department of Energy and Climate Change, *Quarterly Energy Prices: June 2011,* http://www.decc.gov.uk/en/content/cms/statistics/publications/prices/prices.aspx.

34. Consumer Focus, "Jobs, Growth and Warmer Homes: Evaluating the Economic Stimulus of Investing in Energy Efficiency Measures in Fuel Poor Homes," *Cambridge Econometrics, Final Report for Consumer Focus,* October 2012.

35. See EU Fuel Poverty Network, "Working to Raise Awareness of Fuel Poverty."

36. S. Hutton and G. Hardman, *Assessing the Impact of VAT on Fuel on Low-Income Households: Analysis of the Fuel Expenditure Data from the 1991 Family Expenditure Survey* (York: Social Policy Research Unit, University of York, 1993).

37. Gordon Walker, "Decentralised Systems and Fuel Poverty: Are There any Links or Risks?" *Energy Policy* 36 (2008): 4514–4517.

38. Consumer Focus, "Jobs, Growth and Warmer Homes."

39. M. E. Falagas et al., "Seasonality of Mortality: The September Phenomenon in Mediterranean Countries," *Canadian Medical Association Journal* 181 (2009): 484–486.

40. Data for all countries but the UK, see Falagas et al., "Seasonality of Mortality." UK data are from Rosie Day and Russell Hitchings, "'Only Old Ladies Would do That': Age Stigma and Older People's Strategies for Dealing with Winter Cold," *Health and Place* 17 (2011): 885–894.

41. Philippa Howden-Chapman et al., "Tackling Cold Housing and Fuel Poverty in New Zealand: A Review of Policies, Research, and Health Impacts," *Energy Policy* 49 (2012): 134–142.

42. For an extended discussion of this issue, see Tindale and Hewett, "Must the Poor Pay More?"

43. IPPR, "Energy Tax will Hit Business and Undermine European Effort to Tackle Climate Change," April 1, 2013, http://www.ippr.org/press-releases/111/10577/energy-tax-will-hit-business-a-undermine-european-effort-to-tackle-climate-change. For the full IPPR report, see Dominic Maxwell, "Hot Air: The Carbon Price Floor in the UK," June 28, 2011, http://www.ippr.org/publication/55/7629/hot-air-the-carbon-price-floor-in-the-uk.

44. For the downward trend of US electricity prices, see International Energy Agency, *World Energy Outlook 2012,* http://www.worldenergyoutlook.org/publications/weo-2012/.

45. Tindale and Hewett, "Must the Poor Pay More?" 248.

46. Consumer Focus, "Jobs, Growth and Warmer Homes."

47. Walker, "Decentralised Systems and Fuel Poverty," 4517.

48. International Energy Agency, United Nations Development Programme, United Nations Industrial Development Organization, *Energy Poverty.*

49. International Energy Agency, *World Energy Outlook 2012.*

50. See B. K. Sovacool et al., "What Moves and Works: Broadening the Consideration of Energy Poverty," *Energy Policy* 42 (2012): 715–719; and Jamil Masud, Diwesh Sharan, and Bindu N. Lohani, *Energy for All: Addressing the Energy, Environment, and Poverty Nexus in Asia* (Manila: Asian Development Bank, 2007).

51. F. Hussain, "Challenges and Opportunities for Investments in Rural Energy," presentation to the United Nations Economic and Social Commission for Asia and the Pacific (UNESCAP) and International Fund for Agricultural Development (IFAD) Inception Workshop on Leveraging Pro-Poor

Public-Private-Partnerships (5Ps) for Rural Development, United Nations Convention Center, Bangkok, Thailand, September 26, 2011.

52. Masud et al., *Energy for All*.
53. Gwénaëlle Legros, Ines Havet, Nigel Bruce, Sophie Bonjour, Kamal Rijal, Minoru Takada, and Carlos Dora, *The Energy Access Situation in Developing Countries: A Review Focusing on the Least Developed Countries and Sub-Saharan Africa* (New York: World Health Organization and United Nations Development Program, 2009).
54. World Health Organization 2006, 8.
55. Ibid.
56. Masud et al., *Energy for All*.
57. Y. Jin, "Exposure to Indoor Air Pollution from Household Energy Use in Rural China: The Interactions of Technology, Behavior, and Knowledge in Health Risk Management," *Social Science and Medicine* 62 (2006): 3161–3176.
58. S. S. Lim et al., "A Comparative Risk Assessment of Burden of Disease and Injury Attributable to 67 Risk Factors and Risk Factor Clusters in 21 Regions, 1990–2010: A Systematic Analysis for the Global Burden of Disease Study 2010," *Lancet* 380 (2012): 2224–2260.
59. Secondhand cookfire smoke refers to smoke leaving chimneys and dwelling outdoors in populated areas.
60. United Nations Environment Programme, *Natural Selection: Evolving Choices for Renewable Energy Technology and Policy* (New York: United Nations, 2000). Figures have been updated to 2010.
61. Masud et al., *Energy for All*.
62. K. Sangeeta, "Energy Access and Its Implication for Women: A Case Study of Himachal Pradesh, India," presentation to the 31st IAEE International Conference Pre-Conference Workshop on Clean Cooking Fuels, Istanbul, June 2008, 16–17.
63. B. S. Reddy, P. Balachandra, and H.S.K. Nathan, "Universalization of Access to Modern Energy Services in Indian Households: Economic and Policy Analysis," *Energy Policy* 37 (2009): 4645–4657.
64. United Nations Development Programme, "Energy and Major Global Issues," *Energy After Rio: Prospects and Challenges* (New York: United Nations, 1997).
65. Michael Wines, "Toiling in the Dark: Africa's Power Crisis," *The New York Times,* July 29, 2007.
66. William Rafey and Benjamin K. Sovacool, "Competing Discourses of Energy Development: The Implications of the Medupi Coal-Fired Power Plant in South Africa," *Global Environmental Change* 21 (2011): 1141–1151.
67. Ibid.
68. Rafey and Sovacool, "Competing Discourses of Energy Development"; see also, "Dodgy Deal: Medupi Coal Power Plant," *BankTrack*, April 27, 2013, *www.banktrack.org/manage/ajax/ems . . . /medupi_coal_power_plant*.
69. John Vidal, "Britain Has Key Vote on World Bank Loan to Medupi Power Station," *The Guardian,* April 1, 2010.
70. Rafey and Sovacool, "Competing Discourses of Energy Development."
71. Earthlife Africa Jhb, "World Bank's Climate and Governance Disaster," press release, April 9, 2010, http://www.Earthlife.org.za/?p=914.
72. U.S. Energy Information Agency, *South Africa Report,* January 17, 2013, http://www.eia.gov/countries/cab.cfm?fips=SF.

73. "Dodgy Deal" (see note 68).

74. Ibid.

75. "Power Price Rises Will Squeeze Job Priorities," *City Press*, February 3, 2013, http://www.citypress.co.za/business/power-price-rises-will-squeeze-job-priorities/.

76. U.S. Energy Information Agency, *South Africa Report*.

77. "Eskom Consumers Funding BHP's Power," *City Press*, March 23, 2013, http://www.citypress.co.za/business/ eskom-consumers-funding-bhps-power/.

78. "South Africa Crisis Creates Crusading Electricians," *BBC News*, November 24, 2009.

79. City Press, 2013; see also "BHP Billiton 'Expects' Eskom to Respect Power Deals," *The Mail and Guardian* (Reuters), April 3, 2013, http://mg.co.za/article/2013–04–03-bhp-billiton-urges-eskom-to-respect-power-deals.

80. BP Energy Outlook 2030, January 2013, http://www.bp.com/extendedsectiongenericarticle.do? categoryId=9048887&contentId=7082549.

81. Ibid.

82. U.S. Energy Information Administration, "International Energy Statistics: Petroleum Reserves as of 2011," http://www.eia.gov/cfapps/ipdbproject/IEDIndex3.cfm?tid=5&pid=57&aid=6.

83. The only exception among the top five countries with largest oil reserves is Canada. U.S. Energy Information Administration, "International Energy Statistics."

84. Ibid.

85. "The World's 25 Biggest Oil Companies," *Forbes,* January 2013, http://www.forbes.com/pictures/mef45glfe/not-just-the-usual-suspects-2.

86. "Oilfield Services: The Unsung Masters of the Oil Industry," *The Economist*, July 21, 2012, 53.

87. B. K. Sovacool and S. V. Valentine, "Sounding the Alarm: Global Energy Security in the Twenty First Century," in *Energy Security,* ed. B. K. Sovacool (London: Sage, 2013).

88. M. A. Brown and B. K. Sovacool, *Climate Change and Global Energy Security: Technology and Policy Options* (Cambridge, MA: MIT Press, 2011).

89. BP Energy Outlook 2030 (see note 80).

90. Ibid.

91. EIA, "Short-Term Energy Outlook, September 2006," http://www.eia.doe.gov/pub/forecasting/steo/oldsteos/sep06.pdf.

92. International Energy Agency, *World Energy Outlook 2008* (Paris, France: IEA, 2008), Figure 3.10, 105.

93. U.S. Energy Information Administration, *Annual Energy Outlook 2006* (Washington, DC: U.S. Department of Energy, 2006).

94. Larry Shirley, "North Carolina: A State Perspective on Energy," presentation to the North Carolina Department of Administration, April 5, 2006.

95. N. Kumar, "Macroeconomic Overview of the Asia-Pacific Region and Energy Security," presentation to the United Nations Economic and Social Commission for Asia and the Pacific (UNESCAP) Expert Group Meeting on Sustainable Energy Development in Asia and the Pacific, United Nations Convention Center, Bangkok, Thailand, September 27–29, 2006.

96. G. Thapa, "Food, Fuel, and Financial Crises in Asia and the Pacific Region: Impact on the Rural Poor," presentation to the United Nations Economic and Social Commission for Asia and the Pacific (UNESCAP) Expert Group

Meeting on Sustainable Energy Development in Asia and the Pacific, United Nations Convention Center, Bangkok, Thailand, September 27–29, 2006.

97. See CNA, *Powering America's Defense,* as well as Barbara Freese, Steve Clemmer, Claudio Martinez, and Alan Nogee, *A Risky Proposition: The Financial Hazards of New Investments in Coal Plants* (Washington, DC: Union of Concerned Scientists, 2011).

98. Worldwide Fund for Nature, *Coming Clean: The Truth and Future of Coal in the Asia Pacific* (Washington, DC: WWF, 2007).

99. M. Yan, "Volatile Coal Prices Reflect Supply, Demand Uncertainties," *Platt's Insights Magazine* (March 30, 2005): 10–11.

100. Ryan, 2005.

101. International Energy Agency, "Investment in the Coal Industry," background paper on the meeting with the IEA Governing Board, November 2003.

102. International Atomic Energy Agency, *World Distribution of Uranium Deposits (UDEPO) with Uranium Deposit Classification: 2009 Edition,* IAEA-TECDOC-1629 (Vienna: IAEA, 2009).

103. NEA and IAEA, *Projected Costs of Generating Electricity: 2005 Update,* 4345.

5 The sociopolitical dimension

Corruption, authoritarianism, and conflict

Risk exposure is replacing class as the principal inequality of modern society . . .
– Ulrich Beck, "Living in the World Risk Society,"
Economy and Society 35, no. 3 (2006): 333

Introduction

Energy systems both reflect and reinforce the structure of political and economic power within a society, and, as Ulrich Beck suggests with his comment, they distribute risk throughout industrial society. Large-scale, integrated, and centralized electric systems – whether based on fossil fuels or nuclear power – cannot be built or maintained without generous inputs of capital. Operating such systems requires the oversight of some form of centralized authority, and this authority must exercise the power necessary to enforce decisions on controversial issues such as licensing and siting. The extractive industries also tend to be highly concentrated, with a relatively small number of large corporations dominating the international market. In theory, such institutions do not require the same concentrations of wealth and power as the generation and distribution of electricity; the early history of the oil industry in the United States is peppered with "wildcatters" and small family drilling operations. Yet this competitive period did not last long. Today the international oil and gas markets are dominated by those companies in control of extensive financial and technical resources. Given the dependence of conventional energy systems on large amounts of capital, sophisticated technology, and centralized administration, we should not be surprised that the energy industry is often closely linked with authoritarian power structures (even within democratic nations), and tightly coupled mechanisms for control.

The dominant model of energy policy in the United States has long assumed that "a link exists between the level of energy production and the gross national product."[1] This assumption is coupled with another widespread but problematic assumption – one that is recently beginning to receive critical scrutiny – that general welfare increases in direct proportion to GDP. When applied to developing countries, these assumptions lead to the conclusion that investing in conventional energy systems will more or less automatically translate into increased welfare for the citizens of these countries. Increased prosperity will lead to more stable societies with stronger democratic institutions and a more firmly established rule of law.

As an example of this type of thinking, in 1994, at the Asia Pacific Economic Conference in Jakarta, Indonesia, President Bill Clinton signed an agreement to supply Indonesia with electric power using loans from U.S. export credit agencies: the Overseas Private Investment Corporation and the Export Import Bank. The statement he made at the signing reflects the dominant model of U.S. energy policy: "As markets expand, as information flows, the roots of society will grow and strengthen and contribute to stability."[2] In this case, U.S. taxpayer dollars were used for the Paiton power project, which consisted of two 600 MW coal-fired power plants. The power plants were overpriced, mixed up in the corruption of the regime under President Suharto, with the projects being used as vehicles for the enrichment of his immediate family, and eventually contributed to the bankruptcy of the state electric utility, Perusahaan Listrik Negara (PLN).[3]

The decision to build the power plants was forced through by Suharto against the advice of government planners in Indonesia, who had recommended smaller, more distributed gas-fired power plants, pointing to the fact that the "transmission grid leaked like a sieve."[4] Thus, while these coal-fired power projects benefited Suharto and his family circle, and provided billions of dollars of profits for a number of foreign energy corporations, among them the American companies Cal Energy, Mission Energy, and General Electric, it is fair to say that their overall contribution to economic prosperity in Indonesia was negligible, if not negative. From the vantage point of strengthening the rule of law, a central role of the Paiton power project was to serve as a conduit for channeling large amounts of money into undeserving hands.

Centralized energy systems can reinforce authoritarian power structures; they can also undermine democratic forms of governance. The history of the U.S. energy industry demonstrates how difficult it is to balance the many benefits of harnessing the energy of fossil fuels with the dangers to the social order that result from the amassing of power and wealth through the control of energy. The 1890 Sherman Antitrust Act was in large part a response to the abuse of monopoly power exercised by Standard Oil. The concern at the time was that the concentration of social, economic, and political influence exercised by Standard Oil and other trusts was capable of disrupting not only the free enterprise system but the democratic process as well.[5]

As a second example, in 1935 the U.S. Congress passed the Public Utility Holding Company Act (PUHCA) in an attempt to restrain the power and reach of the electric industry, which had been growing through the 1920s and 1930s.[6] The "regulatory compact" worked out between electric utilities and state governments in the early part of the twentieth century, which gave utilities monopolies under the regulatory oversight of state government agencies, had provided a legitimizing framework within which the electric industry could expand its domain. From the 1920s on, the "cozy relationship" between executives and government regulators worked to the advantage of the executives.[7] An investigation by the Federal Trade Commission in 1928, prompted by accusations that utilities had "bought" elections, brought to light the extent of the industry's efforts to manipulate public opinion through propaganda campaigns.[8]

Despite efforts of the New Deal administration to challenge the power of the electric utility system, utility managers maintained their dominance right through to the energy crises of the 1970s. In many ways, the system worked to the benefit of the American public, who enjoyed the cheap electricity that resulted from the economies of scale realized by the utilities. And yet there was little democratic accountability in the decision-making process for this economically vital industry. A large portion of the American public was ignored by the utilities, as they saw little profit to be made from serving rural areas. It took government loans under the 1935 Rural Electrification Administration to help farmers establish cooperative utilities that could serve their needs.

As we will see in this chapter, the Paiton power project, Standard Oil, and utility holding companies are far from isolated examples. Energy systems are not neutral with respect to issues of social and political justice. Conventional energy systems are particularly vulnerable to an abuse of power by those who finance, construct, manage, and administer them, to the point where some have hypothesized the presence of a "resource curse" surrounding oil, gas, and coal.[9] One of the founders of the Organization of Petroleum Exporting Countries (OPEC) even went so far as to call oil "the devil's excrement."[10] The extraordinary profits to be made provide an almost irresistible incentive to consolidate and maintain the dominance of these systems over alternatives, regardless of social costs; moreover, the nature of such integrated and centralized systems requires a great degree of cooperation between energy companies and government authorities, and this affinity lends itself easily to self-interested manipulation by the parties involved, especially as the highly technical nature of energy systems makes democratic accountability difficult. This chapter therefore documents four distinct ways that energy systems impinge on social and political stability: by encouraging corruption, by eroding democratic norms, by violating human rights, and by promoting military conflict.

Energy profits, corruption, and politics

Money rules the world

Analysis of the sociopolitical dimension of energy justice should begin with the most obvious fact of fossil fuel-based energy: It generates huge profits. This money can often be used by companies to bribe politicians, bypass environmental regulations, circumvent democratic procedures, and evade tax payments. Various types of corruption – the looting of oil revenues by well-placed military and political representatives, kickbacks, illegal commissions, and illicit oil-for-arms deals – are routinely associated with revenues from oil exploration and production.[11] In Iran, for example, $11 billion in oil revenues "vanished" over the course of nine months in 2010.[12] In Nigeria, an independent audit recently revealed $540 million "missing" from $1.6 billion in advance payments to oil companies to develop new fields, in addition to 3.1 million barrels of oil that remain unaccounted for, along with the disappearance of $3.8 billion in dividends from natural gas.[13] In Angola, the second largest oil producer in Africa after Nigeria, investigators discovered in December 2000 that $4 billion in oil funds was missing and likely spent illegally on kickbacks to government officials and the purchase of military arms. These missing funds represented at the time a staggering one-third of all state income in Angola.[14] In Congo-Brazzaville, the head of its state-owned oil company, Société nationale des pétroles du Congo, was indicted for illegally taking up to $4.2 million of company revenues per year and depositing them into his personal account.[15] As a senior executive from a multinational oil company told one of the authors, "oil is such a valuable resource that it can be extracted using poor technological and governance techniques and still turn a profit. Oil production covers a multitude of sins."[16]

The corruption associated with oil funds can assume blatant forms in Africa and other parts of the developing world. But the influence of profits from the energy sector is equally pervasive in developed countries, even if this influence is exerted in less direct ways. In 2011, the oil and gas industry spent $150 million lobbying the U.S. Congress; the electric utilities industry, including those companies invested heavily in coal, spent $145 million.[17] It is no secret that the politics of climate change in the United States is heavily influenced by the energy lobby, which is determined to obstruct any attempt to legislate a limit on CO_2 emissions and a shift to alternative sources of energy.

Permissive rules regarding campaign finance provide energy companies with another legal mechanism to subvert the political process. In Europe, attempts to pass a Fuel Quality Directive, which would designate transport fuels from tar sands as 22 percent more carbon intensive than those from convention fuels, have so far been blocked, largely as a result of intense lobbying from the Canadian government, the Canadian Association of Petroleum Producers, and Europe's largest oil companies – including Royal Dutch Shell, British Petroleum, Statoil, and Total – all of which are heavily invested in the Canadian tar sands.[18]

It is not just governments that are at risk of corruption. Even intergovernmental organizations charged with supporting the rule of law and preserving human rights, such as the United Nations (UN), are susceptible. In 2002 and 2003, it was revealed that the United Nations Oil-for-Food Programme – established in 1995 to empower Iraq to sell oil on the world market in exchange for food, medicine, and humanitarian aid despite international sanctions – enabled Saddam Hussein and the Iraqi government to generate $10.1 billion in illegal revenues, including $5.7 billion of smuggling proceeds and $4.4 billion in illicit surcharges on oil products. An independent inquiry by the U.S. Government Accountability Office documented more than 380 illegal contracts, with some of the revenues going directly into the pockets of UN officials managing the program.[19]

Social marginalization and political instability

Financiers, government planners, and energy analysts often perceive energy production as a means to lift countries out of poverty and provide the foreign exchange earnings needed for economic development and social betterment. Yet if we shift our focus from the revenues generated to the effects they have on the social fabric and political stability of the producing country, a very different picture emerges. In Nigeria, for instance, which is the thirteenth largest producer of petroleum in the world, oil and gas accounts for 80 percent of all government revenues, 95 percent of exports, and 90 percent of foreign exchange earnings. Oil production has generated billions of dollars in government revenues, and yet the country's average GDP in the past decade was less than its GDP in the 1960s, and the poverty rate increased from 27 percent in 1980 to 70 percent in 1999.[20] Moreover, rather than bringing peace and stability, oil production has exacerbated political flashpoints. A collection of smaller Niger Delta provinces have attempted to expand their access and control over oil and gas resources, struggling with minority groups clamoring for sovereignty. Militant youth movements have emerged as a result of the social and political insecurity, promoting intensified campaigns of interethnic violence.[21]

Elsewhere in Africa, Chinese firms have embarked upon an "oil safari" with billions of dollars worth of oil contracts signed with Angola, Ethiopia, the Congo, Gabon, Kenya, Madagascar, Nigeria, and the Sudan. These firms have followed a "no-questions-asked" policy of "non-interference," meaning they "don't hold meetings about environmental impact assessments, human rights, bad governance and good governance."[22] In Sudan, the Chinese National Petroleum Corporation (CNPC) funneled $4 billion to the regime in Khartoum. This money enabled the government to purchase Shenyang fighter planes and an assortment of heavy arms, which it then used against the secessionist movement in south Sudan (where many of the oil and gas reserves are located). The CNPC actually permitted the Sudanese government to use its facilities to launch attacks, with groups armed with Chinese weapons.

In Angola, which has extracted some of the world's largest oil and gas reserves for 40 years, average life expectancy had plunged to 36.8 years in 2006, the infant mortality rate had risen to 19 percent, the country imported most of its food, and almost three-quarters of the population lived below the poverty line.[23] As with Nigeria, oil production has thus coincided with a deterioration of the living conditions of the majority of the population. Justice requires an equitable distribution of basic goods among the citizens of a state, so it is fair to ask to what extent oil production has contributed to a fundamentally unjust economic structure. In fact, oil revenues have enabled elites in Angola to capture and exploit virtually all economic enterprises. The proceeds from oil and gas production have nourished oppressive tendencies within the state, which have triggered violent contests for political power from a dissatisfied population. The United Nations has accused Angola of using oil revenues from state-owned corporations to finance blood diamond operations, and the state has admitted to using oil revenues to fund illegal narcotics operations.[24]

The basic pattern in Angola is much the same as in Nigeria and the Sudan: Profits from oil and gas operations, both legal and illegal, lead to an inequitable distribution of winners and losers and the marginalization of a large segment of the population. Social marginalization helps to undermine political stability. In the unsettled and volatile conditions that result from these changes, groups that possess sufficient power attempt to exert control over a greater share of the energy resources of the country. Conflict and war are often the end results, as we will see later in this chapter.

In countries where there is a more firmly established rule of law, the curtailment of civil liberties is sometimes the result of energy projects that benefit elites at the expense of the wider population. In Bangladesh, for instance, the government has ignored civil liberties, such as the freedom of assembly, in its determination to develop an open pit mine (otherwise known as the Phulbari Coal Project) that would extract 15 million tons of coal per year from the Phulbari region, a key rice-producing area. Eight million tons of coal would be exported by rail and barges through the Sundarbans, one of the largest mangrove forests in the world and a classified Biosphere Reserve by the United Nations. Three million tons would be used for domestic consumption; the UK-listed company developing the mine has proposed a 500 MW power plant for the mine site. According to the Private Sector Operations Department of the Asian Development Bank, the project would economically benefit Bangladesh and provide it with much needed energy. The project, however, would involve the relocation of more than 40,000 people from 9,000 households spread across 5,200 hectares of land, and it would also result in water scarcity for 220,000 individuals that would lose access to water supplies as the mine area is de-watered.[25] The project has proceeded without any consultation or consent from the affected communities. When community leaders, environmental groups, and local citizens attempted to peacefully but publicly protest the project in August 2006, state police and paramilitary forces beat thousands of them and fired indiscriminately into the crowd of 50,000, killing three people (including one 14-year-old boy) and injuring more than 100.[26]

Conventional energy systems and authoritarianism

Nuclear power and public participation

One of the requirements of energy justice is that people are provided with the opportunity to participate effectively and meaningfully in decisions concerning the production and distribution of energy, especially when the potential consequences of such decisions could have a detrimental effect on the ability to pursue and acquire the basic goods of human life. Given the potential for catastrophic accidents inherent in the use of nuclear energy, and the ongoing problem of how to dispose of radioactive waste safely, decisions regarding nuclear power ought to involve some degree of public participation. Nonetheless, in the United States, India, France, and the former Soviet Union, nuclear power licensing and siting have all violated the principles of due process and informed consent in various ways.

Legally unable to prohibit the public from participating in administrative procedures, the nuclear industry in the United States has made use of a number of different tactics to minimize public participation in decisions regarding licensing and permitting. Some of the more common practices have included:

- Setting the timing of hearings so that interveners and the public must gather information and present safety concerns before receiving the documents necessary to adequately evaluate the risks of a project
- Interpreting rules narrowly to exclude unfavorable evidence and safety concerns, and
- Attenuating inquiries into safety issues by prematurely ending proceedings.

These practices were designed with the intent to prevent the public from having a significant influence over the siting of nuclear facilities and infrastructure. Public involvement in nuclear hearings was discouraged, and when it did occur the process often alienated the public and served to legitimate nuclear power as a foregone conclusion.[27] MIT nuclear engineering professor David J. Rose once quipped that discussing concerns about nuclear power with members of the nuclear community had "the intelligence, grace, and charity of a duel in the dark with chainsaws."[28]

There is evidence of U.S. nuclear advocates going so far as to manipulate data and public disclosure procedures to serve their own ends. Detailed accident studies carried out by the Atomic Energy Commission (AEC) in the 1950s and 1960s warned of catastrophic adverse consequences associated with a nuclear meltdown, yet public dissemination of these reports was suppressed under pressure from the industry. The AEC functionally "discouraged interest in findings that conflicted with policy optimism." It minimized opportunities for networking among nuclear opponents, limited public awareness of safety concerns, and even suppressed an Advisory Committee on Reactor Safeguards report released in 1956. The results of this report eventually became public in 1973, after repeated requests made by concerned citizens under the Freedom of Information Act.[29]

In one of its most intractable problems – that of radioactive waste – the industry has distorted data and limited any kind of genuine public participation in order to get its way. Studies of consumer attitudes and public opinion have shown that the public will support nuclear power expansion only if assurances of safe waste disposal are provided.[30] For this reason, nuclear power proponents – trade groups, vendors, and utilities – have shifted from a technical discourse, full of uncertainties, to a benevolent discourse of inclusive and respectful public consultation regarding siting issues and criteria of acceptability and safety. Yet one study of such efforts in Canada found that they did not involve an opportunity for genuine consultation, in which citizens had the chance to influence eventual decisions; instead, they were exercises in public relations used to reinvent the industry's image.[31] Nuclear groups employed public consultation sessions to (a) demonstrate consent and approval when they received it, or, in the absence of consent, to (b) portray the public as having fragmented values and conflicting opinions that could never be resolved. The message for regulators was that they should ultimately defer to the nuclear industry in either case. This situation, the study concludes, does not bode well for the democratic process, as the public is coopted either way. Public consultation is converted from a means of informing the goals of public policy into an end in itself.

Nuclear power projects in India have been subjected to expedited environmental impact assessment (EIA) processes so as to minimize public involvement. Moreover, disclosure of information concerning safety and radiation exposure has been left to the discretion of the industry. According to one study assessing the quality of EIAs related to nuclear power facilities:

> Public hearings for nuclear projects have been always short and rushed affairs with insufficient time for all interested participants to seek information or clarifications. . . . Not only is public participation devalued, it often falls short in procedural terms. At a number of hearings, members of the public have complained that they have not managed to get copies of EIA reports. Authorities have not read out their summary (minutes) of the proceedings and sought the consent of those who participated.[32]

To compound the problem, industry representatives are also under no legal obligation to release information regarding the true costs of facilities in addition to health and safety risks.[33]

In the pioneering days of nuclear power, French planners enjoyed the luxury of developing nuclear plans "without any public involvement."[34] When construction of the Superphoenix fast breeder reactor in Creys-Malville began in 1968, the government did not notify citizens of its existence, nor consult local groups, and it disregarded the opinions of those who caught wind of the project. The state simply expropriated land for the project, ignored appeals, and then responded to protests by deploying the Compagnies Républicaines de Sécurité – the special riot police – who killed one demonstrator and wounded 100 others with tear gas and batons.[35] Purdue University professor Daniel P. Aldrich went so far as to call the French government "highly coercive," observing that it relied on "hard social control tools" against opponents of nuclear power. As he put it, "the Ministry of Economy, Finance, and Industry and the Ministry of Interior used police coercion, surveillance, and expropriation to handle anti-nuclear resistance, along with hard social tools to control information flow and access points . . . [and] as a result of the coercion employed by state authorities, the anti-nuclear movement in France has been marginalized."[36]

Disregard for the need to inform the public on critical matters relating to nuclear power was even more pronounced in the former Soviet Union. When the Mayak Industrial Reprocessing Complex in the Southern Urals suffered a devastating accident, the Soviet government evacuated the 272,000 people living around the facility but told them that it was only an exercise. After the Chernobyl accident in 1986, the government did not begin evacuations until April 28 – two full days after the accident – because plant operators had delayed reporting the accident to Moscow out of fear that it would spoil the forthcoming May Day celebrations. Even after learning of the disaster, state officials planned on covering up the full extent of the damage, until a Swedish radiation monitoring station 800 miles northwest of Chernobyl reported abnormally high radiation levels.[37]

Secrecy and democracy

Unlike the hazards associated with burning fossil fuels – the full extent of which have only become apparent to the general public in recent years – the risks of using nuclear power for commercial purposes were evident to most people from the start. For this reason, the imperative to manage the public debate and strictly control flows of information was more pronounced for the nuclear industry. In the past two decades, of course, the fossil fuel industries have also begun to feel the pinch of this imperative, and have accordingly spent billions trying to convince the public that fossil fuels can be used responsibly and that more carbon dioxide in the atmosphere is actually a good thing.[38] In the case of the nuclear industry, it could and has been argued that the highly technical nature of nuclear power precludes the possibility of conveying to the broader public a balanced and adequate understanding of the relevant risks and solutions. Nuclear power is the province of experts. If society at large is to enjoy the benefits of nuclear power – including the fact that it produces little in the way of greenhouse gas emissions – it must be kept in the dark, the thinking goes, about certain technical matters and risks which, if made public, would only be exaggerated and distorted.

It is not our intention here to comment on the advantages or disadvantages of nuclear power. We are interested in the implications of energy production for the analysis of energy justice. In this respect, these examples shine an interesting light on a fundamental characteristic of conventional energy systems that is rarely remarked upon, though some awareness of it is starting to emerge. The history of the nuclear industry demonstrates that authoritarian tendencies are unavoidable in energy systems that are large-scale, highly centralized and integrated, capital-intensive, based on the deployment of sophisticated technology that is beyond the technical understanding of most eventual users of the system, and – as a result of all these features – that require extensive coordination and collaboration between the three dominant power structures within modern societies: industrial, administrative, and political. Authoritarianism, as we are using the phrase, is not primarily a reference to the many autocratic regimes that happen to be major oil producers (though neither is this historical fact merely a coincidence, as we have seen); rather, it is a necessary feature of conventional energy systems, even some large-scale renewable energy systems, and it is therefore operative even in strongly democratic nations. Clearly we need some degree of established and effective lines of coordination and authority if we are to operate something as complex as a national grid – but this should not stop us questioning the extent to which the need for an authoritarian structure within conventional energy systems is compatible with energy justice.

If energy justice demands a maximum of public participation, it would be hard to avoid the conclusion that conventional energy systems by their very nature reinforce and foster unjust sociopolitical structures. For the majority of people, the complex technology and engineering of modern energy systems already present a substantial obstacle to effective participation in decisions regarding energy generation and distribution. However, such an extreme – and unrealistic – conclusion is not required by our analysis. There is a real difference between the subversion of the democratic process characteristic of the nuclear industry in France and the genuine opportunities for public participation provided by a typical rate-making hearing before a state public utility commission in the United States. Energy justice therefore requires proper recognition of the potential for conflict between the right of every person to have some say in decisions affecting his or her environment and the inherently authoritarian nature of conventional energy systems.

Two implications follow from recognition of the potential that energy systems have for institutionalizing an unjust sociopolitical order. First, it is critical that appropriate regulatory safeguards be put into place to ensure that the public has an effective mechanism for making its voice heard. In other words, there needs to be a "countervailing power" to offset the enormous influence wielded by those who control energy systems. Secondly, more thought should be given to alternative energy systems from the perspective of energy justice. Distributed, smaller-scale energy generation that makes use of locally available renewable resources, and does not depend on the extraction of fossil fuels or the mining of uranium, provides greater opportunities for public participation. This latter consideration is especially relevant in the context of the developing world, where in many places energy infrastructure has yet to be built. The choice is between building conventional energy systems that will inevitably marginalize large segments of the population, and developing alternative energy systems based on "appropriate technology," to use the phrase of E. F. Schumacher, that have the potential to enhance the economic wellbeing and political power of people now struggling with poverty and limited opportunities.

Energy and human rights abuses

Human rights abuses go unchecked when there is no one to report them. More often than not, conventional energy resources are located in remote regions populated by indigenous groups or in rural areas with minimal political representation, far from the urban centers where influence and power is concentrated. The institutions of law and order are less visible and effective in the jungle and the desert, and what happens "out there" is not nearly so important as what flows from there into the industrial and financial heartlands: oil, gas, coal, and money.

Multinational energy companies, particularly oil companies, have a deplorable record when it comes to complicity with human rights abuses. In most cases it is not the oil companies themselves that are directly responsible for the abuses. Typically, the abusive actions are carried out under the direction of the project partner from the host country, whether it is the government itself or a local company that has licensed the multinational to extract and transport oil and gas from the region. However, the activities of the oil companies are the occasion for the abuses, and the companies always retain ultimate control over projects, deciding where operations are to take place, how to dispose of waste products, what facilities require protection – decisions which can lead directly to violent abuses of the local population. In far too many cases, oil companies also supply the money, resources, and equipment that are used to suppress local protest and resistance.

As a concern for the question of energy justice, human rights abuses are in a different category than the corruption of political regimes and the undermining of basic civil liberties and democratic procedures. They cut across the categories of civil, cultural, economic, political, and social rights and extend to basic issues of bodily health and integrity, freedom from fear, and the right to enjoy an uncontaminated environment capable of sustaining established patterns of life.[39] Instances of human rights abuse do not require an explanatory framework – description is enough. The stark injustices involved speak for themselves.

Oil and gas companies operating in Indonesia, Myanmar, Nigeria, and Peru have benefited from and at least implicitly sanctioned various human rights abuses, including the suppression of free speech, torture, slavery, forced labor, extrajudicial killings, and executions. Oil companies have consistently employed private security firms to protect their operations and suppress dissent. Royal Dutch Shell gave guns to Nigerian security forces, and Chevron provided aid, helicopters, and pilots to an armed group that then gunned down nonviolent protestors on an oil-drilling platform, killing some and injuring others, who were later tortured. British Petroleum, ExxonMobil, ConocoPhillips, and Royal Dutch Shell continue to provide daily "security briefings" for mercenaries; they also supply vehicles, arms, food, and medicine to soldiers and police.[40]

There is some evidence to suggest that the Talisman Oil Company has been directly involved in the civil war in Afghanistan, with its affiliates possibly deploying mujahedeen and child combatants to provide security around oil and gas blocks.[41] When Talisman was a minority owner in a project developing oil fields in the Sudan, residents of the Sudan brought a lawsuit against the company, alleging that they were victims of genocide and other crimes against humanity. The U.S. court granted summary judgment to Talisman, though the judge was careful to note that her decision by no means implied that the plaintiffs had not suffered great harms and that the government of the Sudan had not committed gross violations of the norms of civilized behavior.[42]

The actions of oil companies at times suggest an attitude of arrogant indifference to humanitarian concerns and questions of basic justice. A comment made to *Newsweek* by a Chevron lobbyist in Washington, D.C., speaking about an environmental lawsuit brought against the company by indigenous people in Ecuador, suggests a remarkable degree of contempt for the developing countries in which Chevron operates: "We can't let little countries screw around with big companies like this."[43]

Resource extraction in unstable regions with limited infrastructure and undeveloped institutions can instigate a scramble for control over resources and intensify already existing ethnic and tribal conflicts, resulting in low intensity warfare or outright civil war (more to come on this point). Human rights abuses often assume truly horrific forms in such situations. Though these abuses are not the sole responsibility of multinational corporations operating in the regions, neither can it be denied that these conflicts and their attendant atrocities are fueled by the profits generated by extracting oil, gas, and minerals. The conflict in the eastern provinces of the Democratic Republic of Congo – a region of great mineral wealth – resulted in the deaths of approximately 5.4 million people between 1998 and 2007, making it the deadliest conflict since World War II.[44] Human rights abuses associated with this conflict include the rape of men, women, girls, boys, and even toddlers; gang rapes committed in front of families and entire communities; male relatives forced at gunpoint to rape their own daughters, mothers, and sisters; women used as sex slaves and forced to eat excrement or the flesh of murdered relatives; young girls murdered by having guns fired into their vaginas; and men forced to simulate sex in holes dug in the ground with razor blades stuck inside.[45]

A few other particular incidents merit elaboration. Royal Dutch Shell paid $15.5 million to settle a lawsuit after it was accused of collaborating with Nigerian militia in the execution of eight activists protesting Shell activities in Nigeria, including Ken Saro-Wiwa and members of a grassroots organization called the Movement for the Survival of the Ogoni People. According to allegations brought by the family of Ken Saro-Wiwa and others, when the group became popular Shell conspired with the militia to capture and hang the men. The plaintiffs also had documents indicating that Shell worked with the Nigerian army to bring about the torture of other protestors and that the company contributed information that resulted in the killing of dissidents. Shell was accused of providing the Nigerian army with vehicles, patrol boats, and ammunition, as well as helping plan raids and military campaigns against villages.[46] Because Shell settled the case, the court did not have the chance to determine the veracity of these allegations.

In 1992, the French oil company Total received a license from Myanmar Oil to develop natural gas from the Yadana field off the coast of Myanmar/Burma. In the same year, the Unocal Corporation, now Chevron, acquired a 28 percent interest in the project from Total. The Myanmar military provided security and other services for the project. In a case brought against Unocal in an American court (which was eventually settled out of court), evidence was offered suggesting that Unocal had known about and implicitly authorized the use of rape, murder, extermination, forced relocation, and slavery to expedite the construction of oil and gas pipelines in Myanmar.[47] In both the Karen and Mon states, the government habitually relied on forced labor to construct pipelines and carried out military attacks on civilians opposing such projects. Unocal admitted in court to knowing that the regime in Myanmar had a record of committing human rights abuses, that the military had forced villagers to work and entire villages to relocate for the purposes of the project, and that the government had committed various acts of violence in front of company employees. A pipeline engineer who has worked on dozens of projects around the world told one of the authors of the present book that "from a human rights perspective, the Yadana and Yetagun pipelines are simply the *worst* because they continue to directly endow a brutal regime with revenue."

Another example involves British Petroleum. The $4 billion Baku-Tbilisi-Ceyhan (BTC) oil pipeline now delivers more than 1 million barrels of oil per day from the Azeri-Chirag-Gunashli fields in the Caspian Sea off the coast of Baku, Azerbaijan, through Georgia, to the Turkish port of Ceyhan on the Mediterranean – traversing 1,768 kilometers. During construction, one Azeri worker was beaten almost to death for questioning a BP manager. A local Azeri NGO, the Oil Workers Rights Protection Committee, awarded some oil executives involved with the BTC project the "torn shoe" award for violating local laws and labor protections, observing that their shoes must be torn from stepping on people all day long. According to labor leaders, the BTC Company and its affiliates refused to hire sufficient local workers, prevented local workers from associating in unions, and failed to provide safe working conditions. The Labor Code of Azerbaijan requires that employers provide employees in dangerous jobs with compulsory insurance and proper training, but in regard to the BTC, the Code was ignored with permission of the government. The Labor Code also requires that companies provide adequate compensation for accidental death or injury, yet seven people who died while working on the pipeline received no compensation. When hundreds of employees working on the BTC orchestrated a series of strikes to protest, they were quickly squashed by the government and some organizers imprisoned.

In recent years, oil and gas companies have gone to great expense to refashion their public images as conscientious global citizens. As the caption for a print advertisement for Chevron shown in Figure 5.1 states, under a photograph of two smiling African women, "Oil Companies Should Support the Communities They're Part Of."[48] But these efforts are little more than public relations campaigns, aimed primarily at Western consumers. In 2008, Chevron defeated a lawsuit in which it was charged with liability for the deaths of protestors on a Nigerian offshore oil platform (the incident referred to above). It then attempted to compel the Nigerian plaintiffs, impoverished widows and children, to reimburse the company for its attorneys' fees. Bert

Figure 5.1 Chevron advertisement proclaiming the company's commitment to local communities, 2010

Source: Chevron.

Voorhees, one of the lawyers for the plaintiffs, commented: "That's how they litigate. . . . The point is to scare off the next community that might try to assert its human rights."[49]

Even when human rights abuses receive publicity – and in most cases they do not, as the instruments of publicity are typically in the hands of those committing the crimes – the legal resources available to the victims are no match for those wielded by multinational oil companies. In most courts of law, if not always in the international court of public opinion, parent corporations can usually distance themselves from the unjust actions of their subsidiaries and affiliates.

Furthermore, it may be impossible to eliminate entirely the inherent human, political, and environmental risks associated with the process of extracting fossil fuels and uranium. As one scholar succinctly put it, "few industrial activities have as large an environmental footprint and are capable of wielding as much influence on the well-being of a society as a large-scale mine or oil and gas project."[50] It is impossible to construct massive open-pit mines or to build thousands of kilometers of pipelines and processing stations without disturbing communities and ecosystems. Such impacts on human wellbeing, on occupational safety, on deforestation, on water quality, and on air pollution can be somewhat "controlled" but "never eliminated." As Keith Slack from Oxfam writes:

> A contradiction thus exists between commitments to operate responsibly and the actual mechanics of how the industry currently functions. Displacing a community of thousands of people in order to dig a massive pit in a pristine mountain or rainforest area, and piling up 300-meter-high mountains of waste rock that will inevitably begin to leach sulfuric acid into groundwater used by local communities will never be seen as socially responsible by some observers.[51]

Slack concludes that for these industries, truly operating in a manner that is responsible to human and ecological needs amounts to a contradiction in terms, a "mission impossible."

Some renewable energy technologies can also involve human rights abuses. Millions of individuals are involuntarily resettled every year due to large-scale hydroelectric projects. About 4 million people are currently displaced by activities relating to hydroelectricity construction or operation annually, and 80 million have been displaced in the past 50 years from the construction of 300 large dams. Yet 31 GW of hydroelectric capacity was added in 2009, an increase in capacity second only to wind power among all sources of renewable energy. Total installed capacity and investments in hydropower dwarfed that of all other major renewable sources of energy in 2009 and 2010; from 2004 to 2009, China roughly doubled its hydroelectric capacity, and significant expansion is expected in Brazil, India, Russia, Turkey, and Vietnam, necessitating the relocation of millions more.[52] This relocation, however, often proceeds without true and meaningful consultation and consent (for more on specific instances, readers should skip ahead to Chapter 6 subsection "Displaced by Energy").

In some cases, it would seem that hydroelectric projects are specifically designed to achieve oppressive political goals, including ethnic cleansing. The Ilisu Dam on the Tigris River, currently under construction in Anatolia in southeastern Turkey, will submerge the ancient city of Hasankeyf when completed. Hasankeyf, whose history reaches back more than 10,000 years, happens to be the cultural capital of the Kurds (against whose nationalist aspirations the Turkish government has been battling for decades), and the hydroelectric project will necessitate the displacement of 78,000 Kurdish residents. Furthermore, by damming the Tigris, Turkey gains leverage over Syria and Iraq, who depend on the waters of the Tigris for meeting basic survival needs.[53]

Energy and military conflict

There is perhaps no greater act of injustice than the aggressive pursuit of war by a state for motives of gain and power. Unfortunately, control over energy resources has often been the motive for armed conflict and war, while the use of high-grade energy has at the same time provided the means for prosecuting wars on an ever more ferocious scale. Our account of the sociopolitical dimension of energy justice ends with a discussion of the close relationship between energy and military conflict.

Armed conflict, terrorism, and civil war

As we discussed in the first section of this chapter, and briefly in the previous section, the extraction and transportation of oil and gas and other mineral resources can have a destabilizing impact on a region or country, bolstering the power of existing elites and regimes and encouraging a scramble to monopolize the profits from these resources. The foreign exchange earnings acquired from oil and gas exports have been used to propagate low-intensity warfare, internal conflict, geopolitical strife, and in some cases, outright war.

Oil revenue has endowed the regime in Khartoum, for example, with the means to expand its military and extend its military campaign in Darfur, Sudan. Oil revenue allowed General Than Schwe to equip the Burmese military with light arms, helicopters, and armored vehicles. Proceeds from the Chad-Cameroon oil pipeline are allegedly helping fund conflicts in the Congo and the Sudan, and the oil and gas money from the Baku-Tbilisi-Ceyhan and South Caucasus Pipelines in Central Asia have enabled Azerbaijan to intensify its military campaign against Armenia.

A protracted civil war in Papua New Guinea was in part the result of the actions of a subsidiary of Rio Tinto, which secreted millions of tons of toxic mine tailings and destroyed more than 30 square kilometers of indigenous rainforest.[54] Many terrorist groups receive funds indirectly from oil and gas revenues and then use those resources to plan attacks against oil and gas infrastructure. A report commissioned by the UN Security Council indicated that from 1993 to 2003 Saudi Arabia used its oil wealth to transfer $500 million to Al Qaeda. In 1997, a high-ranking member of the House of Saud coordinated a $100 million aid package to the Taliban; at the same time, Saudi Arabia was sending hundreds of thousands of barrels of oil a day to Afghanistan and Pakistan in off-budget foreign aid.[55]

These examples are not anecdotal; Shannon O'Leary identified 10 serious civil wars and conflicts from 1990 to 1999 directly fueled by natural resources, most of them energy-related.[56] Michael Ross has also documented numerous cases where oil revenues have exacerbated conflicts around the world.[57] He noted that raising money in petroleum-rich countries can be easy for insurgents and terrorists, who can steal oil and then sell it on the black market, as they did in Iraq and Nigeria. Such groups can also extort money from oil companies working in remote areas, as they did in Colombia and Sudan. Or they can find business partners to fund them in exchange for future consideration if they seize power, which has happened in Equatorial Guinea and the Democratic Republic of the Congo.[58]

In Columbia, oil and gas revenues have contributed to acts of terrorism and intensified military conflict. Guerilla groups reap an oil "war tax" and earn $140 million per year from oil-related extortion and kidnappings. The National Liberation Army and Revolutionary Armed Forces of Columbia, the two largest rebel groups, have extracted war taxes from oil companies and local contractors by threatening sabotage and murders, extortions, and bombings. Because their threats have not always provided them with the leverage they desire, the guerillas have dynamited state oil and gas pipelines more than 1,000 times in 13 years, spilling 2.9 billion barrels of crude oil as a result and greatly damaging local ecosystems and water resources. The economic losses from explosions on the Cano Limon-Covenas pipeline amounted to $1 billion from 1990 to 1995, or 7 percent of Columbia's entire export revenue over the same period. Paramilitary groups have also built a cottage industry by stealing natural gas and gasoline by drilling holes in pipelines. Ecopetrol, a state oil company, estimates they lose $5 million per month. Finally, oil and gas revenues provide the Columbian Army with the significant income needed to purchase weapons and train soldiers. A military tax of roughly $1 per barrel (or $12 to $30 million per year) enables the army to increase its troop presence and prolong its struggle against the rebels.[59]

Geographer Michael J. Watts has gone so far as to argue that oil wealth within the OPEC has largely backed a global arms race, with countries in the Middle East spending roughly $45 billion on arms per year and with every 1 percent increase in oil revenues corresponding to a 3.3 percent increase in arms imports.[60] As he concludes, "the reconfiguration of the global oil industry has produced close alignments between oil, finance, and weapons of war, and it has resulted in a close association between oil security as a strategic concern and various types of conflict."[61]

Energy resources and interstate war

Fossil fuels are rivalrous goods. A barrel of oil in China cannot be used in France. Because of this, conventional energy resources often play a prominent role in major geopolitical rivalries and interstate wars. In World War I, the Entente and Central powers both believed that control of coal, oil, and gas resources was essential to victory. In the decade before the outbreak of World War II, Japan, suffering from a dearth of available raw materials, invaded Manchuria to acquire Manchurian coal reserves. In response to Japan's later invasion of China in 1937, and to show support for the United Kingdom, the Roosevelt Administration abrogated the 1911 Treaty of Commerce and Navigation with Japan in January 1940. As a result, licenses for the export to Japan of gasoline and aviation fuel, as well as machine tools, were suspended. Without domestic resources, Japan invaded the oil-rich Indonesian islands, and the resulting tensions were a contributory factor in Japan's decision to attack Pearl Harbor. That same year, Adolf Hitler declared war on the Soviet Union in part to secure oil for his war machine; he launched *Operation Blau* to protect German oilfields in Romania while securing new ones in the Central Caucasus. The Soviet Union was also concerned about its dependence on foreign oil, and attempted to invade northern Iran in 1945 and 1946 to acquire control of the oil resources there.

Geographer Vaclav Smil has traced the role of energy in war and documented numerous examples where energy reserves prompted or exacerbated international conflict.[62] Apart from identifying the interconnections between energy reserves and the two world wars, Smil argues that energy resources were factors in a number of the major wars of the second half of the twentieth century. In the Korean War, the fact that the northern part of the peninsula contained most of the country's coal deposits was of critical importance. The energy resources at stake in the Vietnam War – waged by France until 1954 and by the United States after 1964 – were oil and gas reserves. The Soviet occupation of Afghanistan in the 1980s was partly about gaining control over the significant energy and mineral resources there. The Gulf War was explicitly about oil, of course, with the Iraqi occupation of Kuwait. Though Smil was researching his article before the second Iraq War, no one can seriously doubt that control over oil reserves in Iraq was a significant part of the American strategy, whatever other valid explanations there might be for the invasion. Smil also suggests that almost all cross-border wars of the late twentieth century were energy-related, including the conflicts between India and Pakistan, Eritrea and Ethiopia, China and India, and also civil wars such as those in Sri Lanka, Uganda, Angola, and Columbia.

The countries of southeast and northeast Asia are also embroiled in intensive interstate rivalries over oil, between themselves and with regional superpowers such as China.[63] To select just a few of the most prominent examples: In 1992, China formally stated its right to the "use of force" to protect its claims to oil and gas resources in the South China Sea; in 1995, China seized a chunk of

Philippine land and attacked a few fishing vessels to procure offshore oil and gas reserves; in the "Kikeh oil dispute," competing efforts between Brunei and Malaysia to conclude contracts with oil companies deteriorated into a tense naval standoff in April and May of 2003.[64] Most recently, in 2013, China and Japan have continued to rattle sabers over oil and gas reserves found within the Spratly Islands, with one Chinese general publicly stating that his country was on the brink of "actual combat" and the usually conservative *Economist* declaring that military conflict remains "a hair-trigger away."[65]

Halfway across the world in South America, energy resources are also a source of border conflicts and political tensions. A proposal in 2002 to export gas from new reserves in Bolivia through a Chilean port triggered grievances long held by Bolivians toward Chile and led to a massive popular uprising. What became known as the "Gas War" left dozens dead and led to the downfall of two Bolivian presidents.[66] Geo-political considerations destabilized the natural gas trade between Venezuela and Colombia; the conflict came to a head in July 2009 when Caracas withdrew its ambassador from Bogota.[67] In Ecuador and Peru, multinational oil companies have partnered with the armed forces to defend their facilities, at times killing protesters on national security grounds.[68] Historically, the bloodiest military conflict in South America in the twentieth century – the Chaco War of 1932–1935 fought between Bolivia and Paraguay – was sparked by speculation regarding petroleum reserves in the Gran Chaco region. Standard Oil and Royal Dutch Shell, each backing a different side, helped propagate the war, their motive apparently to raise oil prices and extend profits.[69]

Dependence on foreign oil is a consideration of critical importance for governments concerned with energy security. Worldwide demand for fossil fuels is quickly rising, and reserves of conventional "easy" oil and gas are shrinking. Even with the non-conventional oil and gas reserves that have recently become accessible through the development of new technology, energy security is a matter of grave national concern across the globe, and has led to increasingly frantic efforts to secure necessary resources. The implications are extremely worrying, as it is quite likely that the twenty-first century will see a sharp escalation in armed conflicts and wars fought over access to fossil fuels. For this reason perhaps more than any other, a proper concern for energy justice requires that we expedite the development and deployment of alternative energy systems.

The technology of modern warfare

Energy is implicated in military action and international conflict in a number of ways. Not only is control over energy resources the source of much armed conflict today, but modern weapons also represent the most concentrated and devastating releases of energy employed by human beings. Furthermore, and partly as a consequence of the nature of modern weapons, military operations require the mobilization of energy resources on a vast scale, to such an

Table 5.1 Kinetic energy released by weapons and explosives

Weapon	Projectile/explosive	Kinetic energy (J)
Bow and arrow	Arrow	20
Heavy crossbow	Arrow	100
Civil War musket	Bullet	1×10^3
M16 assault rifle	Bullet	2×10^3
Medieval cannon	Stone ball	50×10^3
Eighteenth century cannon	Iron ball	300×10^3
World War I artillery gun	Shrapnel shell	1×10^6
Hand grenade	TNT	2×10^6
World War II heavy AA gun	High-explosive shell	6×10^6
M1A1 Abrams tank	Depleted uranium shell	6×10^6
Unguided World War II rocket	Missile with payload	18×10^6
Suicide bomber	TDX	100×10^6
500 kg truck bomb	ANFO	2×10^9
Boeing 767 (September 11, 2001)	Hijacked plane	4×10^9
Hiroshima atomic bomb (1945)	Fission	52×10^{12}
U.S. nuclear intercontinental ballistic missile	Fusion	1×10^{15}
Novaya Zemlya bomb (1961)	Fusion	240×10^{15}

Source: Vaclav Smil.

extent that research into energy efficiency and renewable sources of energy has become one of the top priorities for military planners today.[70]

Energy is embodied in weapons and projectiles, which release all of it at once to inflict a maximum amount of damage. As Table 5.1 reveals, the history of weaponry can also be regarded as a history of the development of the human ability to store and release destructive energy. Some forms of weaponry, such as depleted uranium shells, utilize byproducts exclusively from energy activities, such as uranium mining and milling.

The construction of weaponry with this magnitude of destructive power requires large inputs of energy and the use of energy-intensive materials. Once built, these technologies require significant flows of liquid fuels and electricity to function. At a "conservative estimate," Smil calculates that about 5 percent of all U.S. and Soviet commercial energy consumed between 1950 and 1990 went to developing and amassing weapons and their delivery mechanisms.[71] As another indication of the energy-intensive nature of modern warfare, it is notable that the American military now risks most of its combat causalities from convoys and their guards, primarily distributing inefficiently used fuel.[72] By one estimate, 50 percent of the energy used by the U.S. Air Force is spent hauling energy fuels.[73]

The second law of thermodynamics, sometimes known as the law of entropy, states that every time energy is converted from one form to another, the availability of that energy to perform useful work is reduced.[74] Fossil fuels have provided mankind with an invaluable store of accessible high-grade, low entropy energy. What we are consuming when we burn coal, oil, and gas is the order inherent in them;[75] from the perspective of the laws of thermodynamics, our energy consumption represents an intensification of the irreversible process of universal entropic degradation. In other words, fossil fuel consumption is a process of converting order into chaos, transforming useful energy into dispersed energy that is no longer accessible for human use. When fossil fuels are used to perform work that increases human wellbeing, a case can be made that the costs of increasing thermodynamic disorder (and thereby irrevocably reducing the supply of accessible energy) is balanced by the benefits of creating a flourishing human order. But this application of cost-benefit analysis makes no sense when energy is used to destroy the material basis of human societies. Given the finite and exhaustible nature of easily accessed forms of high-grade energy, and the negative consequences of burning carbon for present and future generations, the use of concentrated energy for purely destructive purposes is a particularly egregious example of energy injustice.

Conclusion

The many cases of energy injustice we have considered in this chapter all violate in some form or another the prohibitive principle of energy justice. One of the primary justifications for any sociopolitical system worth defending is that it ensures that the people living within it have secure access to those goods that are necessary for human prosperity, from physical wellbeing and adequate dwelling to basic civil and political rights. As we have seen in this chapter, conventional energy systems often undermine social and political stability. Every phase of the global energy system – from extraction and transportation to generation and distribution – has the potential (if not regulated carefully) to strengthen the disruptive and exploitative forces within society and to interfere with the ability of communities and governments to provide for a just distribution of goods.

We do not intend to minimize the many social benefits brought about by conventional energy systems, and we recognize that fossil fuels and nuclear power will be part of the energy portfolio for many decades to come. However, if governments and energy decision-makers continue to ignore the realities we discuss in this chapter (and summarize below), we have no doubt that the sociopolitical injustices associated with energy systems will only increase as conventional energy resources become more scarce and the international competition for them becomes more fierce.

- The huge profits gained from oil and gas production often corrupt the political process, in both developing and developed nations.
- Resource extraction and energy production can strengthen the power of autocratic regimes and ruling elites, exacerbating existing inequalities,

marginalizing portions of the population, and undermining social cohesion and stability.

- Because the development of oil and gas fields, pipelines, hydroelectric projects, and coal mines often requires the displacement of large numbers of people, governments wishing to pursue such projects often choose to ignore the civil and political rights of the people so displaced.
- Foreign exchange revenues from the sale of oil and gas are frequently used by autocratic governments to purchase military hardware that is then employed against political opponents and minority ethnic groups.
- Centralized integrated conventional energy systems have a tendency to foster the development of authoritarian institutional structures that sideline and at times subvert democratic participation in decisions of broad public interest.
- The power and wealth that can be acquired through fossil fuel extraction, combined with the remote locations of many oil and gas fields, provide powerful incentives for human rights abuses.
- Attempts to control access to energy resources provoke armed conflict at the local, regional, and international levels. As energy security becomes of paramount concern to governments throughout the world, with diminishing resources and growing global demand, the likelihood of global energy conflicts and war increases substantially.

Notes

1. Joseph P. Tomain and Richard D. Cudahy, *Energy Law,* 2nd ed. (St. Paul: Thomson Reuters, 2011), 103; see also Richard F. Hirsh, *Power Loss: The Origins of Deregulation and Restructuring in the American Electric Utility System* (Cambridge, MA: MIT Press, 1999), 50–51.
2. "Deceit and Deception: US Dealings in Indonesia," *Scoop.co.nz,* September 24, 1999.
3. "How US Companies and Suharto's Circle Electrified Indonesia," *The Wall Street Journal,* December 23, 1998.
4. Ibid. See also, Environmental Defense Fund's International Case Study, *Indonesian Power Projects: The Paiton Debacle* (2004), www.edf.org/documents/2445_case study_indonesianpower.pdf (accessed April 8, 2013).
5. Robert H. Lande, "Wealth Transfers as the Original and Primary Concern of Antitrust: The Efficiency Interpretation Challenged," *Hastings Law Journal* 34 (1982): 83, 96–101.
6. Tomain and Cudahy, *Energy Law*, 372–373.
7. Ibid., 45.
8. Hirsh, *Power Loss*, 40.
9. Macartan Humphreys, Jeffrey D. Sachs, and Joseph E. Stiglitz, "What Is the Problem with Natural Resource Wealth?" *Escaping the Resource Curse* (New York: Columbia University Press), 1–20.

10. B. K. Sovacool, "The Political Economy of Oil and Gas in Southeast Asia: Heading Towards the Natural Resource Curse?" *Pacific Review* 23, no. 2 (2010): 225–259.

11. Michael J. Watts, "Righteous Oil: Human Rights, the Oil Complex, and Corporate Social Responsibility," *Annual Review of Environment and Resources* 30 (2005): 373–407.

12. Cyrus Maximus, "Islamic Republic Corruption Scandal: $11 Billion in Oil Money Missing," March 28, 2011, http://iranchannel.org/archives/962.

13. Joe Brock and Tim Cocks, "Nigeria Oil Corruption Highlighted by Audits," Reuters News Service, March 8, 2012.

14. Shari Bryan and Barrie Hofmann, *Transparency and Accountability in Africa's Extractive Industries: The Role of the Legislature* (Washington, DC: National Democratic Institute for International Affairs, 2007), 36–37.

15. Ibid.

16. Sovacool, "The Political Economy of Oil and Gas."

17. Center for Responsive Politics, "Lobbying Database," *OpenSecrets.org*, April 2013, http://www.opensecrets.org/lobby/.

18. "EU Tar Sands Pollution Vote Ends in Deadlock," *The Guardian,* February 23, 2012.

19. Joseph A. Christoff, "Observations on the Oil for Food Program," testimony before the Committee on Foreign Relations, U.S. Senate, Washington, DC, US GAO, April 7, 2004.

20. Yinka O. Omorogbe, "Alternative Regulation and Governance Reform in Resource-Rich Developing Countries of Africa," in *Regulating Energy and Natural Resources,* ed. Barry Barton, Lila Barrera-Hernandez, Alastair Lucas, and Anita Ronne (Oxford: Oxford University Press, 2006), 39–65.

21. Michael Watts, "Resource Curse? Governmentality, Oil, and Power in the Niger Delta, Nigeria," *Geopolitics* 9, no. 1 (2004): 50–69.

22. Ian Taylor, "China's Oil Diplomacy in Africa," *International Affairs* 82, no. 5 (2006): 937–959.

23. Omorogbe, "Alternative Regulation and Governance Reform."

24. Richard M. Auty, "Natural Resources and Civil Strife: A Two-Stage Process," *Geopolitics* 9, no. 1 (2004): 29–48.

25. Bank Information Center, "ADB's Private Sector Arm Considering Funding Controversial Coal Project in Northwest Bangladesh," April 2012, http://www.bicusa.org/en/Project.59.aspx.

26. International Accountability Project, "Open Letter to Financial Institutions Investing in Global Coal Management Regarding the Phulbari Coal Project," Bangladesh, August 2008.

27. Steven L. Del Sesto, *Science, Politics, and Controversy: Civilian Nuclear Power in the United States, 1946–1974* (Boulder, CO: Westview Press, 1979).

28. David J. Rose, "Energy and History," *American Heritage* 32 (1981): 79–80.

29. Steven Mark Cohn, *Too Cheap to Meter: An Economic and Philosophical Analysis of the Nuclear Dream* (New York: State University of New York Press, 1997).

30. See Frans Berkhout, *Radioactive Waste: Politics and Technology* (London: Routledge, 1991); A. Blowers, D. Lowry, and B. Solomon, *The International Politics of Nuclear Waste* (New York: MacMillan, 1991); A. Blowers, "Nuclear Waste and Landscapes of Risk," *Landscapes Research* 24, no. 3 (1999): 241–264; and Peter

Stoett, "Toward Renewed Legitimacy? Nuclear Power, Global Warming, and Security," *Global Environmental Politics* 3, no. 1 (2003): 99–116.

31. Darrin Durant, "Buying Globally, Acting Locally: Control and Co-Option in Nuclear Waste Management," *Science and Public Policy* 34, no. 7 (2007): 515–528.

32. M. V. Ramana and Divya Badami Rao, "The Environmental Impact Assessment Process for Nuclear Facilities: An Examination of the Indian Experience," *Environmental Impact Assessment Review* 30 (2010): 268–271.

33. Ibid.

34. Charles de Saillan, "Disposal of Spent Nuclear Fuel in the United States and Europe: A Persistent Environmental Problem," *Harvard Environmental Law Review* 34 (2010): 462–519.

35. Daniel P. Aldrich, *Site Fights: Divisive Facilities and Civil Society in Japan and the West* (Ithaca, NY: Cornell University Press, 2008).

36. Ibid.

37. B. K. Sovacool, *Contesting the Future of Nuclear Power: A Critical Global Assessment of Atomic Energy* (London: World Scientific, 2011).

38. See Ross Gelbspan, *Boiling Point: How Politicians, Big Oil and Coal, Journalists, and Activists Have Fueled a Climate Crisis – And What We Can Do to Avert Disaster* (New York: Basic Books, 2005); Riley E. Dunlap and Aaron M. McCright, "Climate Change Denial: Sources, Actors and Strategies," in *Routledge Handbook of Climate Change and Society,* ed. Constance Lever-Tracy (London: Routledge, 2010), 240–259; John J. Berger, *Climate Myths: The Campaign against Climate Science* (Berkeley, CA: Northbrae Books, 2013).

39. Chris Ballard and Glenn Banks, "Resource Wars: The Anthropology of Mining," *Annual Review of Anthropology* 32 (2003): 305.

40. B. K. Sovacool, *The Dirty Energy Dilemma* (Westport, CT: Praegar, 2008), 134.

41. Watts, "Righteous Oil."

42. See *Presbyterian Church of Sudan v. Talisman Energy, Inc.,* 453 F. Su2d 633 (S.D.N.Y. 2006).

43. Patrick Radden Keefe, "Reversal of Fortune," *The New Yorker,* January 9, 2012.

44. See International Rescue Committee (IRC), *Mortality in the Democratic Republic of Congo: An Ongoing Crisis,* 2007, http://www.rescue.org/news/irc-study-shows-congos-neglected-crisis-leaves-54-million-dead-peace-deal-n-kivu-increased-aid – 4331.

45. "Atrocities Beyond Words," *The Economist,* May 3, 2008.

46. "Shell Pays Out $15.5m Over Saro-Wiwa Killing," *The Guardian,* June 9, 2009.

47. B. K. Sovacool, "Reassessing Energy Security and the Trans-ASEAN Natural Gas Pipeline Network in Southeast Asia," *Pacific Affairs* 82, no. 3 (2009): 467–486.

48. Keefe, "Reversal of Fortune."

49. Ibid.

50. Gavin Hilson, "Corporate Social Responsibility in the Extractive Industries: Experiences from Developing Countries," *Resources Policy* 37 (2012): 131–137.

51. Keith Slack, "Mission Impossible?: Adopting a CSR-Based Business Model for Extractive Industries in Developing Countries," *Resources Policy* 37 (2012): 179–184.

52. B. K. Sovacool and L. C. Bulan, "Behind an Ambitious Megaproject in Asia: The History and Implications of the Bakun Hydroelectric Dam in Borneo," *Energy Policy* 39, no. 9 (2011): 4842–4859.

53. George Monbiot, "They're All Dammed," *The Guardian,* February 26, 2002.
54. Robert Paehlke, "Environmental Harm and Corporate Crime," in *Corporate Crime: Contemporary Debates,* ed. Frank Pearce and Laureen Snider (Toronto: University of Toronto Press, 1992), 305–306.
55. Robert Baer, "The Fall of the House of Saud," *Atlantic Monthly,* May 2003, 34–48.
56. Shannon O'Leary, "Resources and Conflict in the Caspian Sea," *Geopolitics* 9, no. 1 (2004): 161–175.
57. Michael L. Ross, "Blood Barrels: Why Oil Wealth Fuels Conflict," *Foreign Affairs* 87, no. 3 (2008): 2–8.
58. Ibid., 2.
59. Thad Dunning and Leslie Wirpsa, "Oil and the Political Economy of Conflict in Columbia and Beyond: A Linkages Approach," *Geopolitics* 9, no. 1 (2004): 81–92.
60. Watts, "Righteous Oil."
61. Ibid., 378.
62. Vaclav Smil, "War and Energy," in *Encyclopedia of Energy,* vol. 2, ed. Cutler Cleveland (New York: Elsevier, 2004), 363–371.
63. Andrew T. H. Tan, "The Security of the ASEAN Energy Supply Chain," presented at the Seminar on Sustainable Development and Energy Security, Institute of Southeast Asian Studies, Singapore, April 22–23, 2008.
64. Sovacool, "Reassessing Energy Security."
65. "The Senkaku/Diaoyu Islands: Dangerous Shoals," *The Economist*, January 19, 2013, 12–13.
66. Patricia I. Vasquez, "Energy and Conflicts: A Growing Concern in Latin America," Energy Working Paper, an Inter-American Dialogue report, November 2010.
67. Ibid.
68. Erika Gonzalez, Kristina Saez, and Pedro Ramiro, "Multi-National Companies and the Energy Crisis in Latin America," in *Sparking a Worldwide Energy Revolution: Social Struggles in the Transition to a Post-Petrol World,* ed. Koyla Abramsky (Oakland, CA: AK Press, 2010), 178–187.
69. Marc Gavalda, "Recuperating the Gas: Bolivia in its Labyrinth," in *Sparking a Worldwide Energy Revolution: Social Struggles in the Transition to a Post-Petrol World,* ed. Koyla Abramsky (Oakland, CA: AK Press, 2010), 208–218.
70. See U.S. Department of Defense, *DOD Renewable Energy Assessment,* DOD Report to Congress, Washington, DC, March 14, 2005; Get W. Moy, "Reducing the Vulnerabilities of Department of Defense Utilities and Energy Use," in *Solutions for Energy Security & Facility Management Challenges,* ed. Joyce Wells (Lilburn, GA: The Fairmont Press, 2002); Larry C. Triola, "Energy and National Security: An Exploration of Threats, Solutions, and Alternative Futures," *IEEE Energy 2030,* November 17–18, 2008, 13–35; Claudette Roulo, "Clean Energy Tied to National Security," American Forces Press Service, February 7, 2013, http://www.defense.gov/news/newsarticle.aspx?id=119237; and John Daly, "U.S Military Gets Serious about Biofuels," March 26, 2012, http://oilprice.com/Alternative-Energy/Biofuels/U.S.-Military-gets-Serious-about-Biofuels.html.
71. Smil, "War and Energy."

72. Amory Lovins, preface to the Chinese edition of *Winning the Oil Endgame* (Snowmass, CO: Rocky Mountain Institute, 2009).
73. Interview with Michal Quah, Energy Studies Institute, Singapore, June 14, 2009.
74. Nicholas Georgescu-Roegen, "Energy and Economic Myths," in *Valuing the Earth: Economics, Ecology, Ethics,* ed. Herman E. Daly and Kenneth N. Townsend (Cambridge, MA: MIT Press, 1993), 89–112.
75. Jonathon Porritt, *Capitalism as if the World Mattered* (London: Earthscan, 2005).

6   The geographic dimension

Uneven development and
environmental risks

The bourgeoisie has only one solution to its pollution problems: it moves them around.
— Frederick Engels, *The Condition of the Working Class in
England* (Manchester, 1845), 44

Introduction

We live in a world of statistical averages. Average income per capita, average consumption per household, average literacy rate – these are undoubtedly useful categories that give us some idea of how we are performing as a society. However, these averages become deceptive when lumped together and presented as *national* averages. For example, it would be unwise to expect the income of an average Russian citizen residing within Moscow's Garden Ring to be even close to the national average of $18,945 when a pint of beer goes for $8 to $10 at a local pub.[1] Unsurprisingly, in the late 2000s, Moscow's GDP per capita was 19 times that of Russia's poorest region, the Republic of Ingushetia.[2]

The geographic distribution of economic prosperity within a nation state is often staggering. The problem is not unique to developed or developing countries, newly independent or long-time sovereigns – virtually no nation is immune. A glance at NASA's image of Earth, presented in Figure 6.1, confirms the notion that the distribution of prosperity brought by access to energy does not follow geopolitical borders.[3] Although the difference between developed and developing countries is visible, the difference within individual countries is profound. Yet even NASA's image does not capture the geographic distribution of burdens and externalities generated by the world's energy system.

In this chapter, we go beyond national borders – and NASA maps – to examine the role of energy in the allocation of energy risks. In our inquiry, we range from a region comprising several large territorial units, such as provinces and states, to neighborhoods within municipalities. We ultimately conclude, somewhat similar to Frederick Engel's observation, that the global energy system largely addresses its problem of pollution by "moving it around." That is,

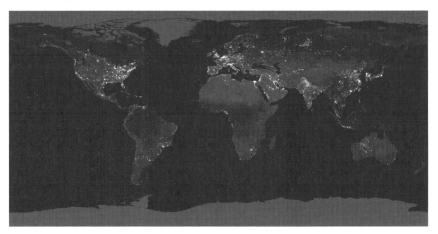

Figure 6.1 The earth at night, 2012

Source: U.S. National Aeronautics and Space Administration and National Oceanic and Atmospheric Administration. Composite map of the world assembled from data acquired by the Suomi NPP satellite in April and October 2012.

the current energy system contributes to an inequitable geographic distribution of economic benefits, as well as an unbalanced distribution of environmental and socioeconomic costs.

Uneven economic development

Beyond oversimplified dichotomies

Global South and Global North. The developing world and the developed word. Poor countries and rich countries. These are clear and straightforward sets of terms that each paint the world in just two colors. Such a simple approach proves useful for energy outlooks, where energy demand, supply, and other metrics are divided between the OECD and non-OECD countries. Unfortunately, this degree of generalization often carries over to climate negotiations and energy policy discourse, where under the UN Framework Convention on Climate Change we have the division of "industrialized" (Annex I) countries and developing (non-Annex I) countries.[4]

However, the view of regions and countries as homogenous economic spaces is highly misleading. As just one example, a yacht owner in India who lives in a million-dollar house will consume far more energy and emit more greenhouse gases than a rural farmer in Tennessee. Even within a country, as we show in this chapter, the economic status of a part-time, hourly wage construction worker in Mississippi, one of the states comprising America's Deep South, is quite different than that of his counterpart in Chicago or New York. Similarly, the salary of a waitress at a centrally located restaurant in Luanda could cause her counterpart in inner Angola to question whether the two live and work in the same country.

The current structure of the world's energy system is partially to blame, for it contributes to geographically uneven economic development in myriad ways. First, centralized power generation makes economic sense if it is located close to load centers. With few exceptions, these load centers are densely populated urban areas.[5] As a result, outlying areas and remote communities often do not get the same degree of reliable and affordable energy access. This translates into fewer economic opportunities for local businesses, higher costs for heating and lighting of schools, and lower levels of medical and other social services. Second, exploration, development, and extraction of fossil fuel, mineral, and uranium deposits, because they are depletable, lead to short and often uncontrolled spikes in economic activity in some locales, followed by prolonged periods of economic depression. When construction of a pipeline ends or an oil field has run dry, jobs evaporate, leaving behind a disappointed younger generation, diminishing municipal budgets, and infrastructure that will rust and decay for many years to come.

Remarkably, the problem of "ghost" mining towns is visible on every continent aside from Antarctica. The landscapes of Ordos in Inner Mongolia, China; Pripyat in Ukraine; Bothwellhaugh in North Lanarkshire, United Kingdom; Ballara in Queensland, Australia; Kolmanskop in Namibia; Centralia in Pennsylvania, the United States; and Sewell in Chile, serve as reminders of economic promise turned into regional economic deceleration. Third, some energy projects, such as large-hydroelectric dams, mountaintop removal, and valley fill operations (for coal in particular), eliminate local economic activities altogether by erasing the space in which these activities took place. Local communities, often indigenous peoples, are forced to relocate, leaving behind the area that served as the foundation of what energy planners would call their "primitive" economy. Often overlooked is the fact that the displaced community had sustained the "primitive" economy for many generations, from long before the advent of the electric light bulb or Watt's steam engine.

Where transmission lines end

The reason behind the uneven geographic reach of modern electricity networks is succinctly captured in the World Bank's report on rural electrification:

> Support for electrification has mostly been provided to communities where connection was deemed most cost-effective, leaving remote communities – often among the poorest – the last ones connected. This pattern is at best only partially overcome by the development of off-grid electricity sources, but per unit costs of off-grid sources are significantly higher than the price of electricity on the grid.[6]

Simply put, siting, building, running, and decommissioning large generation facilities in what energy executives and planners see as the "middle of
nowhere" is cost-prohibitive. When power plants do not have urban masses
to serve, extending transmission lines to small towns, villages, and individual
homesteads is practically tantamount to "financial suicide." The laws of physics make it even harder for outlying communities to get connected, as the percentage of power delivered to the end user diminishes with every additional
leg of a transmission or distribution line (or, in the case of hydrocarbons, the
more pressure you need to push oil or gas through a pipe). Reducing the size
of most "conventional" generation units is not a solution, either, as it goes
against the biggest advantage of conventional power generation – economies
of scale. That is why diesel generators, the conventional go-to solution for
off-grid power generation, are not cost effective and therefore only rarely supported by the World Bank.[7]

Conventional nuclear power plants are an even worse choice for rural
electrification as their financial success is premised on generating base load
power in massive quantities. As noted above, sparsely populated areas lack
the demand for such power, making this marginally cost-effective technology
even less attractive.[8] It is worth noting that proponents of nuclear energy have
high hopes for Small Modular Reactors (SMRs) that can be manufactured
at a factory and "plugged in" to a remote area in need of electric power.[9]
According to some experts, the modularity of such reactors may replace the
conventional nuclear power plants' economies of scale.[10] At this point, it is
unknown whether SMRs will be economically competitive, as the industry
is yet to release their production costs. Should SMRs succeed, their proliferation will signal only a partial departure from the centralized model. The
units, and therefore outlying communities, will depend on a centralized and
highly regulated nuclear fuel cycle system.[11]

The results of this structural deficiency (concerning the extent and
reach of modern energy networks) are highly visible. Currently, more than
3.6 billion people either do not have access to electricity or are completely
dependent on traditional fuels (such as firewood, charcoal, dung, and hard
coal) for their cooking and thermal needs.[12] Of these, almost 83 percent
(3 billion people) live in rural areas.[13] In certain parts of the world, the percentage is even higher, reaching up to 91 percent in sub-Saharan Africa.[14]
The impact on communities is severe, as businesses are slow to grow and
education and medical services are often unavailable. For example, in
Ghana, an electrified rural community is 50 percent more likely to have a
health facility and secondary school, 15 times more likely to have a post
office, and 21 times more likely to have a municipal water system than a
non-electrified one.[15]

The struggles of rural communities in the developing world to overcome
the barriers of their current energy systems are in line with what rural dwellers of the United States had to wrestle with in the 1930s. By the beginning
of that decade, 90 percent of the country's urban population had access to

electric power, while 90 percent of the rural population did not.[16] As a result, 5 out of 6 American farmers had to milk their cows in the dark by hand.[17] It took a monumental effort, including establishment of the Rural Electrification Administration and enactment of the Rural Electrification Act, part of Franklin D. Roosevelt's New Deal, to bring electric power to rural America on a wide scale.

Up, but mostly down

Energy companies in the business of exploring, developing, extracting, and transporting fossil fuels, or mining and leeching uranium, often point to economic benefits that the underlying activities bring to local communities. Given the size of the industry and the scale of activities, the *potential* for positive impact certainly exists. Consider the global mining industry. It consists of roughly 550 large corporations operating more than 20,000 mines spread across the world, employing more than 1 million miners, extracting more than 40 billion tons of raw product valued at $962 billion per year,[18] to say nothing of the informal mining sector that employs an additional 13 million people worldwide, affecting the livelihood of an estimated 80 to 100 million people.[19] Coal accounted for about 7 of the 40 billion tons, and uranium about 45,000 tons in 2010 (low in volume, but ostensibly high in value).[20]

   Unfortunately, in many cases, mining companies do little to realize the full potential for economic development. In fact, some go as far as to make the economic status of communities worse. Consider the examples of coercion in Australia. Uranium mining at the Kakadu National Park in the Northern Territory has a long history of utilizing coercive tactics to override opposition. Operators of both the Jabiluka Mine and the Ranger Mine have been documented intimidating, illegally imprisoning, bullying, and bribing the indigenous Mirrar people into signing over land rights. One academic assessment of these practices in Australia concluded that the mining industry has taken a "devil may care" attitude toward the impacts of its operations: operating in areas without social legitimacy, causing major devastation, and then leaving when an area has been exhausted of all economically valuable resources."[21]

   In cases where prosperity rolls into town with drilling rigs and mining equipment, the wealth stays only for as long as oil, natural gas, coal, and other minerals are being developed and extracted. In fact, it is rare to encounter sustained economic growth in an area known mainly for its extractive activities. The norm, unfortunately, is an economic boom followed by a prolonged economic bust.[22] Luckily, such an unsustainable economic development curvature does not take all communities by surprise. For example, Chris Johnson, the mayor of Elko, Nevada, chose to run his booming town "like it could crash at any time using cash, not much borrowing."[23] And if he ever deserves a reminder of the consequences of an economic bust, he only needs to travel a short distance to nearby abandoned mining towns.[24]

According to Daryl Dukart, Commissioner of Dunn County in North Dakota, not everyone in the 3,800-strong community has a positive attitude toward the economic boom brought by the Bakken shale development.[25] It is not difficult to comprehend why. Home prices have gone up drastically.[26] For example, homes priced at $50,000 in 2006 were selling for $175,000 in 2012.[27] There are many more cars and trucks on the road, evidenced by an almost 400 percent increase in traffic violations and 1,600 percent increase in costs of road maintenance between 2006 and 2010.[28] Dunn County's social and emergency services are extremely stressed and, despite the apparent prosperity, fuel assistance requests have increased.[29] Thus, despite the influx of jobs and business opportunities, a number of local residents are frightened and would like to "go back to what it was like before oil development."[30]

Unfortunately, not *all* or even *most* of the apparent wealth that is supposed to converge on the community actually comes and stays there. For instance, in Russia, the Tyumen region (including the Khanty-Mansi and Yamal-Nenets Autonomous Districts) is responsible for 66 percent of the nation's oil and 91 percent of the nation's natural gas production.[31] This is a colossal contribution, considering that over 60 percent of Russia's export revenues come from oil and gas sales.[32] Yet, according to the official statistics, the region provides only 12 percent of Russia's industrial production and the same percentage of Gross Regional Product (GRP).[33] In contrast, Moscow accounts for 23 percent, the largest share of Russia's GRP.[34] A substantial part of this figure comes from fossil fuels extraction. Readers who have visited the Russian capital may find these numbers perplexing, as no drilling rigs, quarries, or transmission pipelines are visible within the highly packed megalopolis. The answer to this paradox lies in the practice called transfer pricing that allows extraction companies to allocate their profits to Moscow, where their headquarters are located. As such, 10 percent of all Russia's energy production and fossil fuel extraction is statistically "allocated" to Moscow, bringing a huge windfall to the city budget and taking funds from the regions that paid the environmental and socioeconomic price.[35]

Community displacement

The previous two subsections covered the instances of energy injustice *contributing* to geographically uneven economic development. This subsection examines situations where a community is uprooted and its economic base *destroyed* by an energy project. Consider the following examples from around the world involving many different energy systems.

In Sarawak, the largest state of Malaysia, the Sarawak Corridor of Renewable Energy would build no less than 12 hydroelectric dams connected to industrial facilities along the coast of Borneo. The Corridor would extend for some 320 kilometers from Tanjung Manis to Samalajau, covering an area of 70,709 square kilometers, more than half the size of the state. The master plan calls for $105 billion worth of investment by 2030 with the goal of expanding the Sarawak economy by a factor of 5, increasing the number of jobs in the state by a factor of 2.5, and doubling the population to 4.6 million.

A series of legitimate questions arises in connection with such a master plan, however. How do the planners accommodate the existence of the subsistence economy on which the Sarawak indigenous population has relied, and, most importantly, in many cases, on which it wants to continue relying? What value are environmentally sustainable agricultural, hunting, and gathering practices given under the conventional GDP and GRP analyses? Are picked berries, roasted wild boar, and fresh fish caught with a spear taken into account in calculating the purchasing (or, in this case, bartering) power parity of an economy that will grow by a factor of five?

Unfortunately, the subsistence economy is usually an afterthought of the planning process.[36] The environmental impact assessment associated with the first of these dams, the Bakun Hydroelectric Project, was conducted only after construction commenced, and has been attacked for vastly underestimating the facility's negative effect on water quality, ecosystems, and communities. The Sarawak "master planners" have not only disregarded the value of the subsistence economy, they also have refused to recognize the methods that indigenous communities use to establish native customary land tenure. Key decisions to proceed with projects like Bakun and Murum also occurred without the consent of and meaningful consultation with affected communities, even though the Bakun dam necessitated the forceful removal of 10,000 indigenous people and Murum will require the resettlement of 3,400 people. Community leaders in opposition to the project are reputed to have been beaten, tortured, and, in some cases, murdered.[37] The much-touted economic benefits of these two projects never reached the people who sacrificed the most. The Sarawak indigenous communities lost the place that has been their home, lumberyard, and breadbasket for many generations. In return, they did not even receive access to electricity, the generation of which was the sole reason for their removal. Rather than providing electric power to local communities, the dams will instead supply all of their electricity to factories and smelters.

A community is not always directly coerced to leave the area. Sometimes an energy project makes the area unsuitable for a certain type of economic activity and the community is forced, indirectly, to relocate. Halfway across the world, Turkey, Syria, and Iran have all heavily dammed the headwaters that flow into the Tigris and Euphrates Rivers for electricity and irrigation. The combined result is much less water for Iraq, leaving the Shatt al Arab River without enough supply for livestock, crops, and drinking. Tens of thousands of Iraqi farmers have already had to abandon their fields and homes for lack of water.[38]

Similarly, in China and the Greater Mekong Subregion of Asia, under its last five-year state plan China failed to meet its hydroelectric targets and is now playing catch-up. Its most recent master plan calls for hydropower to meet roughly 15 percent of its national energy needs. On the eight great

Tibetan rivers alone, almost 20 dams have been built or are under construction while some 40 more are planned or proposed.[39] For the Mekong River Delta specifically, China's Great Water Diversion Plan involves the upper reaches of the Brahmaputra, Mekong, Salween, and Arun Rivers, where it intends to build 11 dams on the Lower Mekong River alone and 77 more dams throughout these other rivers. However, these sources of water are home to 60 million people that depend on the river system for their food, water, and livelihoods. One study calculated that the fishery losses associated with these dams, which the Chinese will not pay for, will amount to 26 to 42 percent loss of catch per year, more than $476 million of economic damage.[40] Another study calculated even worse economic carnage, with $2 to $3 billion in yearly losses derived from losses of 2.5 million tons of fish.[41] With nothing left to sustain them, the residents of thousands of fishing communities spread across Cambodia, Laos, and Vietnam will likely become refugees.

Although dams are a type of energy infrastructure that unequivocally forces people to leave their homelands, construction of oil and gas pipelines induces relocation as well. In many cases, like that of Southeast Asia, it is an indigenous population that is forced to move. Such resettlement is rarely accompanied by meaningful and adequate compensation. As extraction and transport commence, adverse health issues usually become concentrated among the poor and those living near the pipeline, including toxic leaks, gas flares, and dumping that contaminate food and drinking sources. Finally, gas production often inflates the prices of local goods and services, further hurting the poor outlying communities. Instead of creating corridors of prosperity, pipelines contribute to degradation of neighboring communities. For instance, proximity to a pipeline often magnifies rates of prostitution and sexually transmitted diseases such as HIV, and draws labor away from traditional sources.[42]

The construction of the Baku Tbilisi Ceyhan pipeline (BTC) destroyed roads, small-scale wells and water distribution systems, farming land, and community buildings. What British Petroleum and the BTC Company termed as "consultation" with villagers amounted to little more than officials from these companies giving lengthy presentations and instructions followed by one two-minute question-and-answer session (with some villages restricted to only one question). Resettlement action plans were supposed to be in place 60 days before construction in each village, so that people would be informed about land acquisition and resettlement, but in practice some villagers were never notified at all. As a result, many Azerbaijanis and Georgians protested the project by lying down in front of bulldozers and filing thousands of complaints and 37 lawsuits. In some cases, such as the village of Zayam in the Shemkir Region of Azerbaijan, communities received no consultation at all, and villagers claimed that a BTC Company employee falsified their documents and kept the money for himself.[43]

The BTC project has consequently harmed local communities. One extensive independent study conducted after all sections of the pipeline were built in Azerbaijan involved research visits to 86 communities along the pipeline corridor, interviews of 3,000 people, and the collection of 600 questionnaires.[44] It found that more than three-quarters (76 percent) of those surveyed were unhappy with the BTC Project. The report noted that 7,500 internally displaced people were still living in tents and unlikely to relocate. It estimated that 90 percent of funds provided by the BTC Company and others for social relocation and community development went instead to foreign NGOs. It noted that roads in the Shamkir, Goranboy, Ujar, and Kurdamir regions sustained significant damage from construction but had not been repaired. It documented that at least 50 homes endured structural damage but received no compensation. And it found that many irrigation systems sustained damage from construction but were also never repaired.

However, it is perhaps the mining industry that takes "first prize" in the displacement of communities. One wide-ranging international survey estimated that at least 2.6 million people have been displaced due to mining in India from 1950 to 2009,[45] and individual mines in Brazil, Ghana, Indonesia, and South Africa have involuntary displaced between 15,000 to 37,000 people from their homes.[46] That study noted two troubling conclusions. First, the impact of such displacement extends beyond loss of land, which represented only 10 to 20 percent of impoverishment impacts, to include joblessness, homelessness, marginalization, food insecurity, increased health risks, social disarticulation, and the loss of civil and human rights. Second, it found that compensation for resettlement rarely occurred, and that when it did, such compensation was insufficient to restore – let alone improve – quality of life for the displaced.

As described above, often little or no legitimate and equitable institutional framework exists to facilitate the relocation process. A dam, pipeline, strip mine, or nuclear power plant is deemed to be in the national interest, the interest of *the people* of the nation. Such a determination is based on a broad policy statement, principal, or even a constitutional provision that, when applied, contradicts the rules that *the people* have lived by for generations. Moreover, entire legal frameworks can be restructured in the name of *the people*. As a result, communities that are full of *people* are either deprived of the means of their existence (e.g., land) or the means of their existence suffer from degradation (e.g., water supply or fishing rights).

In Indonesia, state-owned coal mining enterprises, responsible for most of the nation's energy production, are poorly regulated and have extensively displaced communities without consultation or consent. Part of this involuntary resettlement has to do with Article 33 of Indonesia's constitution, which states that the natural resources of the country are to be exploited under state control – with no need for informed consent – for the maximum benefit of the people of Indonesia. The Ombilin coalmine, with reserves of 109 million tons, has disregarded local land rights entirely and improperly managed acid rock drainage and sediment ponds to the degree that 60 square kilometers of land had to be abandoned by indigenous communities.[47]

In China, the state's policies on land acquisition, lack of judicial independence, and low levels of public environmental awareness help fast-track the siting of nuclear plants and subordinate attempts to stop facilities from being constructed. For much of the past century, the Chinese constitution actually banned private land transactions; meaning land was "neither considered as a commodity nor as an asset for producing economic wealth."[48] Instead, it belonged entirely to the state so that land could be allocated more efficiently and agricultural productivity would be accelerated.[49] Even after widespread reforms in the 1970s, property rights and land ownership still remained "collective" and dictated completely by government officials.[50] This enabled the state to distribute parcels of land for nuclear power plants and related facilities directly and with complete control, meaning no contest exists over nuclear power siting, and the state retains the ability to relocate any settlements that hinder nuclear construction projects. In essence, these land uses and acquisition policies make legal challenges to nuclear power plants "impossible."[51]

These examples put disregard for property rights and rearrangement of legal and regulatory regimes in the interest of *the people* into question. What politicians, CEOs, and bureaucrats fail to recognize is the fact that *the people* are comprised of communities, and when communities die, part of *the people* perishes as well.

The point of the previous discussion is not to suggest that centralized power generation facilities and fossil fuels have absolutely no place in the world's energy system. We fully recognize the value of economies of scale in powering large cities and the necessity of certain fossil fuels, natural gas in particular, in making a transition to a lower-carbon energy future. Neither do we suggest that all the economic problems confronting rural communities arise from the current energy system. Political regimes, ethnic tensions, and many other factors contribute to geographically uneven economic development. However, the current energy system has, is, and, unfortunately, will continue to play a role in creating economic inequality. Therefore, to create a more equitable future, nations must take into account the instances of injustice highlighted above during the decision-making process.

Uneven distribution of externalities and risks

In this section, we turn to institutions, procedures, and processes that result in the unequal distribution of existing and potential (i.e., risks) energy externalities. We draw predominately from examples found in developed countries that showcase erosion of the vitality of communities surrounding energy facilities.

Three worlds in one country

Many forms of energy production promote "environmental racism" or the "milk and cream" problem: both within and between communities, a certain proportion of the population reaps the benefits of cleaner sources of energy while another, often larger, proportion becomes worse off, being left disorganized and polluted.[52] Dirty infrastructure can sometimes create "national sacrifice zones" that condemn poorer communities to suffer disproportionately.[53] Most of those working in energy-related jobs with greater occupational hazards (such as coal mines or refineries) tend to be near the poverty line; poorer families have less capital to invest in energy efficiency and thus live in homes that consume more electricity; and lower income families live in neighborhoods in closer proximity to conventional power plants, high voltage transmission lines, nuclear reactors, municipal landfills, trash incinerators, pipelines, abandoned toxic dumps, and nuclear waste repositories, exposing these families to the life-endangering pollution that these plants bring.[54]

One study concluded that communities in the vicinity of nuclear power and waste sites, as well as dumps, mines, and other energy facilities, express characteristics that make them "peripheral communities," set apart from vibrant (and often wealthier) urban communities.[55] These peripheral communities tend to be:

- *remote,* either geographically separated from population centers or relatively inaccessible;
- *economically marginal,* with most being homogenous in terms of social and demographic background and dependent on the nuclear industry as a dominant employer;
- *politically powerless,* with most key political decisions being made elsewhere, often in metropolitan centers;
- *culturally defensive,* with residents expressing ambivalent or ambiguous attitudes toward nuclear energy, combined with feelings of isolation and a fatalistic acceptance of nuclear activities; and
- *environmentally degraded,* meaning they tend to occupy previously polluted land or are close to places where radioactive risks are already present.

In essence, dirty or unwanted energy facilities will invariably migrate to communities that lack the political, social, and economic strength to oppose them, especially indigenous peoples and tribes, often at the extreme social and geographical periphery of society.[56]

Communities comprised of racial minorities must bear a disproportionate share of the world's noxious risks and environmental hazards, as the consequences of energy production move "from white, affluent suburbs to neighborhoods of those without clout."[57] One meta-analysis of studies documenting the spatial distribution of pollution found "clear and unequivocal evidence that income and racial biases in the distribution of environmental hazards exist."[58] A similar assessment of environmental pollution across 2,083 counties in the United States found that "toxic releases increase as a function of the [minority] population."[59]

Consider the following example from the United States. More than two-thirds of all African Americans live within 30 miles of a coal-fired power plant. They are rushed to the emergency room for asthma attacks at more than four times the national average, and have children three times as likely to be hospitalized for treatment of asthma.[60] African Americans consume more fish in larger portions than other Americans, meaning that they have a higher exposure to mercury poisoning from power plants.[61] Consequently, about half of African American children have unacceptable levels of mercury and lead in their bloodstream compared to 16 percent of the general population, and nationwide studies demonstrate that the air in communities of color contain higher levels of ambient particulate matter, carbon monoxide, ozone, and sulfur dioxide.[62]

In Eastern Europe, the Roma have been displaced from so many countries and cities that they are forced to reside in settlements akin to "environmental time bombs." Roma communities in the Czech Republic and Slovakia, for example, reside in flats located above abandoned mines where they are vulnerable to flooding and susceptible to breathing methane gas. Others live in abandoned factory sites surrounded by mining waste where children are fully exposed to toxins and suffer long-term health effects.[63]

As has been repeatedly noted, energy burdens and risks are especially unkind to the poor. As a result, disadvantaged communities and regions bear a disproportionate share of energy-related externalities. An argument can be made that some blue-collar workers make a conscious choice to overlook certain environmental hazards in exchange for economic stability. However, most companies have moved their operations overseas and left many environmental hazards in place. The factory might be closed, but the power plant is still in full operation, sending electricity to towns and cities far away. Ohio, the heart of not only the nation but also America's Rust Belt, has the most polluted air of any state in the nation. The 1.4 million residents of the state's Cuyahoga County face a cancer risk more than 100 times the limit established by the Clean Air Act.[64] One of the most comprehensive studies ever undertaken on environmental externalities, a $3 million, three-year study by the Oak Ridge National Laboratory and Resources for the Future, found that power plant pollution was primarily responsible for increased mortality among the elderly, the very young, and individuals with preexisting respiratory disease.[65]

Landfills in Scotland and nuclear waste storage facilities in Taiwan have also followed a similar trend, with studies confirming that the poor suffer the "triple jeopardy" of being exposed to higher levels of pollution and more likely to suffer health impacts, while being least responsible for generating air quality problems in the first place.[66] These examples reveal a three-world nation, region, state, and even city, where instances of energy injustice are prevalent at multiple geographic scales.

Climate refugees

We conclude the chapter by noting the geographic distribution of the impacts from global climate change, the greatest environmental challenge the world has ever faced. Its impending threats have already pushed, and will continue pushing, many families out of their homes, communities out of their villages and towns, and entire nations out of their borders. These climate refugees, or environmental refugees, must relocate due to rising sea levels, extreme weather events, drought, water scarcity, and other climate-related developments. According to the Environmental Justice Foundation, "[e]very year climate change is attributable for the deaths of over [hundreds of thousands of] people, seriously affects a further 325 million people, and causes economic losses of $125 billion."[67] A separate study calculated that by 2050 more than 200 million people could lose their homes due to climate change.[68]

While calamities have been common throughout the ages, world population growth worsens the scope of the environmental refugee problem. As the New York Times explains, "[W]ith the prospect of worsening climate conditions over the next few decades, experts on migration say tens of millions more people in the developing world could be on the move because of disasters."[69] Similarly, Tuvalu, the Marshall Islands, Papua New Guinea, and low-lying parts of the Caribbean could be submerged within 60 years if sea levels continue to rise. The Republic of Kiribati, a small island country in the Pacific, has already had to relocate to higher ground 94,000 people living in shoreline communities and coral atolls.[70] The Republic of the Maldives could lose 80 percent of its land due to rises in sea level and has already started purchasing land in Sri Lanka for its "climate refugees."[71]

Some countries will be more affected than others. In Bangladesh, coastal areas comprise some 32 percent of the country's total area, and more than 35 million people live in coastal areas less than one meter above sea level. Several studies indicate that the vulnerability of the coastal zone to climatic changes could worsen in the near term due to the confluence of sea level rises, subsidence, changes of upstream river discharges, cyclones, and the erosion of coastal embankments. Severe saltwater intrusion of drinking water supplies will occur, and natural disasters will become both more frequent and intense.

Earthen embankments constructed by the Bangladesh Water Development Board will be subject to accelerated erosion, and with a 45-centimeter rise of sea level, 15 percent of the land in Bangladesh will likely be inundated by the year 2050, resulting in more than 25 million climate refugees from the coastal districts.[72] Bangladesh needs to invest approximately $4 billion per year, now and every year, in adaptation measures such as embankments and cyclone shelters; tragically, the *entire* government budget is roughly only $10 billion per year. Even though Bangladesh will suffer "in the worst way" from climate change, its carbon dioxide emissions are one-twentieth the global average and about one-hundredth the level of U.S. emissions.[73]

Interestingly, the geographic distribution of climate change's effects will not necessarily follow the pattern of other environmental externalities.[74] Global climate change does not discriminate: Low-lying wealthy countries like the Netherlands and affluent communities like Manhattan have been and are certain to be further impacted. The Dutch know that the sea level is rising and have a sophisticated National Programme for Spatial Adaptation to Climate Change.[75] New Yorkers had a preview of things to come after Hurricane Sandy loosened its grip on lower Manhattan:

> Ordinarily, veteran New Yorkers don't like to be caught looking up. But when a power center loses its power, it's no ordinary tourist sight. Residents stared open-mouthed Tuesday at the darkened, sodden outline of the Manhattan financial district left by Hurricane Sandy. A triangular shard of glass hung, shivering, from a blown-out office window, revealing an empty brown desk. The steps of the mighty JPMorgan Chase building looked like a muddy shore. Strings of stoplights swung in the wind, dead as fish eyes. In the thick-stoned Wall Street area, buildings that a day ago seemed impregnable now looked pathetically defenseless, their vestibules and mezzanines turned into deep swimming holes.[76]

All in all, Hurricane Sandy caused up to $50 billion in damages to New York (and that's excluding the destruction in its wake as it traveled across the Bahamas, Cuba, Dominican Republic, Haiti, and Puerto Rico, and in other U.S. states such as New Jersey).[77] In essence, the United States saw a climate change-related disaster influence the outcome of the 2012 U.S. presidential election and reinstate the topic of climate change within the American political discourse.[78] Hopefully, the images of devastation suffered by the coastal communities in New York and New Jersey will continue to serve as a reminder that while justice should always strive to be blind, the injustice arising from climate change will almost certainly be blind as well.

Conclusion

There is no doubt that energy affects, and sometimes determines, the development of a nation state. Some countries, such as Russia and Saudi Arabia, find their national identity closely affiliated with energy production. However, what gets lost in the world of national averages, geopolitics, and economic ambitions are the real people connected to places they call home. The countless stories of marginalized neighborhoods, towns, and villages demonstrate how lives can be adversely affected by energy production, transportation, and use. A conflict exists between the current energy system and the prohibitive and affirmative energy justice principles.

People, and the spaces where they live, matter – regardless of their proximity to a capital city or corporate headquarters. Ignorance of this fundamental rule leads to injustice. This is the reason why we conclude with the following observations that we hope energy planers, politicians, and corporate decision-makers take into account while devising energy policies and contemplating energy projects.

- Conventional electric power generation and distribution systems tend to be structurally prohibitive for equitable geographic development.
- Fossil fuel production in outlying areas is typically characterized by sharp spikes in economic activity, followed by prolonged periods of stagnation and decay.
- When temporary oil-, natural gas-, and coal-generated wealth comes into a rural area, a significant portion of it bypasses local residents as it moves to central cities where extraction companies are headquartered.
- People, especially indigenous populations, who are displaced by energy projects lose their livelihoods and, consequently, struggle to find their way back to being economically productive and self-sufficient members of society.
- Geographic allocation of energy-related externalities usually does not follow allocation of economic benefits, making poor communities even poorer, segregated communities even less racially diverse, and struggling states even more economically stagnated.
- Rising sea levels, increasing extreme-weather events, and other climate change-related externalities may become the biggest harbingers of energy injustice, as people who contributed least to the problem are more likely to lose their homes, towns, and even countries.

Notes

1. Tim Gosling, "Russia Tops BRIC Country List Economically," Russia Beyond the Headlines, October 6, 2010, http://www.telegraph.co.uk/sponsored/rus sianow/business/8046011/Russia-tops-Bric-country-list-economically.html.
2. Federal Service for State Statistics, "Russian Regions," 2011, http://www.gks.ru/bgd/regl/B11_14p/IssWWW.exe/Stg/d01/11–02.htm.
3. NASA, Earth Observatory, http://earthobservatory.nasa.gov/NaturalHazards/view.php?id = 79765 (accessed May 8, 2013).
4. United Nations, "Framework Convention on Climate Change, Parties, and Observers," http://unfccc.int/parties_and_observers/items/2704.php (accessed May 8, 2013).
5. One notable exception would be Norilsk, Russia, the home of the world's largest nickel and palladium processing plant. Norilsk is located above the Arctic Circle in Western Siberia, Russia. The city of approximately 105,000 inhabitants is home to Norilsk Nickel, the largest mining company in Russia. Norilsk Nickel, "Fact Sheet," http://www.nornik.ru/en/investor/fact/ (accessed May 8, 2013). According to Rosstat, the Russian statistics agency, Norilsk is the most polluted city in the country. RIA Novosti, Po Dannym Rosstata, Samym Gryaznym Gorodom Rosii Stal Norilsk [According to the Rosstat, Norilsk is the Dirtiest City in Russia], June 22, 2011, http://ria.ru/nature/20110622/391764826.html.
6. The World Bank, Independent Evaluation Group, "The Welfare Impact of Rural Electrification: A Reassessment of the Costs and Benefits: An IEG Impact Evaluation at 56," 2008 (hereafter, the World Bank Report).
7. Ibid.

8. See U.S. Energy Information Administration, "Levelized Cost of New Generation Resources in the Annual Energy Outlook," 2013, http://www.eia.gov/forecasts/aeo/electricity_generation.cfm, showing relatively (in comparison to other conventional technologies) high LCOE.

9. See the Center for Strategic and International Studies video, *An Emerging Technology: Small Modular Reactors*, September 13, 2012, http://csis.org/multimedia/video-emerging-technology-small-modular-reactors.

10. Ibid.

11. We expand on the technological shortcomings of nuclear power in the next chapter.

12. World Bank, "Rural Electrification Funds: Sample Operational Documents and Resources," http://pworldbank.org/public-private-partnership/sector/energy/laws-regulations/rural-electrification-funds.

13. Ibid.

14. Ibid.

15. World Bank, "Impact of Rural Electrification on Microenterprise," http://web.worldbank.org/WBSITE/EXTERNAL/EXTOED/EXTRURELECT/0,,contentMDK:21610930~menuPK:4489096~pagePK:64829573~piPK:64829550~theSitePK:4489015~isCURL:Y,00.html.

16. University of Wisconsin Center for Cooperatives, "Rural Electric," http://reic.uwcc.wisc.edu/electric/.

17. Middle Tennessee Electric Membership Corporation, "History of Electric Cooperatives," http://www.mtemc.com/acrobat/coop_history_lowrez.pdf.

18. "Global Stock Values Top $50 Trillion: Industry Data," Reuters, March 21, 2007, http://www.reuters.com/article/2007/03/21/us-markets-shares-values-idUSL2144839620070321.

19. Gavin Bridge, "Contested Terrain: Mining and the Environment," *Annual Review of Environment and Resources* 29 (2004): 205–259.

20. "Global Mining Industry Overview," *MBendi,* 2010, http://www.mbendi.com/indy/ming/p0005.html.

21. Heledd Jenkins, "Corporate Social Responsibility and the Mining Industry: Conflicts and Constructs," *Corporate Social Responsibility and Environmental Management* 11 (2004): 23–34.

22. Examples of extraction "booms" and "busts" are so plentiful we thought that highlighting a few would not do the justice to the size and scale of the problem. See generally, Richard M. Auty, *Sustaining Development in Mineral Economies: The Resource Curse Thesis* (New York: Routlege, 2003); Terry L. Karl, "Oil-Led Development: Social, Political, and Economic Consequences," Working Paper No. 80, Center on Democracy, Development, and the Rule of Law, Stanford University, January 2007. See also Bridge, "Contested Terrain," and "Global Mining Industry Overview."

23. "Boom Town, U.S.A.," *Planet Money,* National Public Radio, November 23, 2011, available at http://www.npr.org/templates/transcript/transcript.php?storyId=142675977.

24. Ibid.

25. Daryl Dukart, *Dunn County North Dakota's Tight Oil Development Center for Strategic and International Studies,* presentation at Developing North America's Unconventional Oil Resources: Focus on Tight Oil, Center for Strategic and International Studies, November 12, 2007, http://csis.org/files/attachments/111207_EnergyDukartPPT.pdf.

26. Ibid.
27. Ibid.
28. Ibid.
29. Ibid.
30. Ibid.
31. U.N. Development Programme, *National Human Development Report for the Russian Federation 2009 Energy Sector and Sustainable Development* 28 (2010), http://hdr.undp.org/en/reports/nationalreports/europethecis/russia/name,20196,en.html (hereafter, *UNDP Report*).
32. Keun-Wook Paik, *Sino-Russian Oil and Gas Cooperation: The Reality and Implications,* vol. 30 (Oxford: Oxford Institute for Energy Studies, 2012).
33. Ibid.; *UNDP Report*, 28.
34. Ibid.
35. Ibid.
36. This problem is not unique to Malaysia. The Canadian government, for example, does not even include subsistence economy in Gross National Product. Such an omission is surprising because the value of food harvested by Nunavut is at least tantamount to that brought from Southern parts of the country. Larry Simpson, "The Subsistence Economy," http://www.nunavut.com/nunavut99/english/subsistence.html.
37. B. K. Sovacool and L. C. Bulan, "Energy Security and Hydropower Development in Malaysia: The Drivers and Challenges Facing the Sarawak Corridor of Renewable Energy (SCORE)," *Renewable Energy* 40, no. 1 (2012): 113–129; B. K. Sovacool and L. C. Bulan, "Meeting Targets, Missing People: The Energy Security Implications of the Sarawak Corridor of Renewable Energy (SCORE) in Malaysia," *Contemporary Southeast Asia* 33, no. 1 (2011): 56–82; and B. K. Sovacool and L. C. Bulan, "They'll Be Dammed: The Sustainability Implications of the Sarawak Corridor of Renewable Energy (SCORE) in Malaysia," *Sustainability Science* 8, no. 1 (2013): 121–133.
38. Steven Lee Myers, "Lament for a Once-Lovely Waterway," *The New York Times*, June 12, 2010; Steven Lee Myers, "Vital River is Withering, and Iraq Has No Answer," *The New York Times*, June 12, 2010.
39. Denis D. Gray, "China's Dams Threaten Livelihoods," *The Star,* April 26, 2011.
40. Hezri Adnan, "Riveted by River Resources," *New Straits Times*, December 4, 2012, 16.
41. Francois Molle, Tira Foran, and Mira Kiikonen, *Contested Waterscapes in the Mekong Region Hydropower, Livelihoods and Governance* (London: Earthscan, 2009), 228.
42. Sovacool Pacific Affairs 2009. The same criticism has also been levied at the global mining industry, known for the strongly patriarchal nature of its work force and high rates of alcoholism, prostitution, violence, and HIV rates. Chris Ballard and Glenn Banks, "Resource Wars: The Anthropology of Mining," *Annual Review of Anthropology* 32 (2003): 287–313. One study estimated that between 20 and 30 percent of miners in South Africa were HIV positive. Bridge, "Contested Terrain."
43. B. K. Sovacool, "Cursed by Crude: The Corporatist Resource Curse and the Baku-Tbilisi-Ceyhan (BTC) Pipeline," *Environmental Policy and Governance* 21, no. 1 (2011): 42–57; B. K. Sovacool, "Reconfiguring Territoriality and Energy Security: Global Production Networks and the Baku-Tbilisi-Ceyhan (BTC) Pipeline," *Journal of Cleaner Production* 32, no. 9 (2012): 210–218.

44. G. Guliyeva, I. Huseynli, M. Gahramanly, Z. Ismayilov, S. Ramazanov, and T. Eminli, *Monitoring Working Group on Assessment of Social Impacts of BTC Pipeline During The Construction Phase* (Baku: National NGO Monitoring of Azerbaijan Section of Baku–Tbilisi–Ceyhan [BTC] Oil Pipeline, 2005).

45. Theodore E. Downing, *Avoiding New Poverty: Mining-Inducted Displacement and Resettlement* (International Institute for Environment and Development, 2002).

46. Ibid.

47. Chris Ballard, *Human Rights and the Mining Sector in Indonesia: A Baseline Study* (International Institute for Environment and Development, 2001).

48. Chengri Ding, "Land Policy Reform in China: Assessment and Prospects," *Land Use Policy* 20 (2003): 109–120.

49. See Xiao-Yuan Dong, "Two-Tier Land Tenure System and Sustained Economic Growth in Post-1978 Rural China," *World Development* 24, no. 5 (1996): 915–928; and Samuel P. S. Ho and George C. S. Lin, "Emerging Land Markets in Rural and Urban China: Policies and Practices," *The China Quarterly* 175 (2003): 681–707.

50. Loren Brandt, Jikun Huang, Guo Li, and Scott Rozelle, "Land Rights in Rural China: Facts, Fictions and Issues," *The China Journal* 47 (2002): 67–97.

51. Chi-Jen Yang, "A Comparison of the Nuclear Options for Greenhouse Gas Mitigation in China and in the United States," *Energy Policy* 39, no. 6 (2011): 3025–3028.

52. Elise Boulding and Kenneth E. Boulding, *The Future: Images and Processes* (London: Sage Publications, 1995).

53. Robert D. Bullard, *Unequal Protection: Environmental Justice and Communities of Color* (San Francisco: Sierra Club Books, 1994).

54. Dorothy K. Newman and Don Day, *The American Energy Consumer* (Cambridge, MA: Ballinger Publishing Company, 1975).

55. A. Blowers and P. Leroy, "Power, Politics, and Environmental Inequality: A Theoretical and Empirical Analysis of the Process of 'Peripheralisation,'" *Environmental Politics* 3, no. 2 (1994): 197–228.

56. C. Michael Rasmussen, "Getting Access to Billions of Dollars and Having a Nuclear Waste Backyard," *Journal of Land Resources and Environmental Law* 18 (1998): 335–367.

57. Kimberlianne Podlas, "A New Sword to Slay the Dragon: Using New York Law to Combat Environmental Racism," *Fordham Urban Law Journal* 23 (1996): 1283–1294.

58. Paul Mohai and Bunyan Bryant, *Environmental Racism: Reviewing the Evidence* (Boulder, CO: Westview Press, 1992), 174.

59. David W. Allen, "Social Class, Race, and Toxic Releases in American Counties," *The Social Science Journal* 38 (2001): 13–25.

60. Deanne M. Ottaviano, *Environmental Justice: New Clean Air Act Regulations and the Anticipated Impact on Minority Communities* (New York: Lawyer's Committee for Civil Rights Under Law, 2003).

61. Martha H. Keating and Felicia Davis, *Air of Injustice: African Americans and Power Plant Pollution* (Washington, DC: Clean the Air Task Force, 2002).

62. Adam Swartz, "Environmental Justice: A Survey of the Ailments of Environmental Racism," *The Social Justice Law Review* 2 (1994): 35–37.

63. Tamara Steger, *Making the Case for Environmental Justice in Central and Eastern Europe* (Budapest: CEU Center for Environmental Law and Policy, 2007).

64. Ottaviano, *Environmental Justice*, 5–8.

65. U.S. Department of Energy and the Commission of the European Communities, *U.S.-EC Fuel Cycle Study: Background Document to the Approach and Issues,* ORNL/M-2500 (Knoxville, TN: Oak Ridge National Laboratory, 1992); U.S. Department of Energy and the Commission of the European Communities, *Estimating Externalities of Coal Fuel Cycles,* UDI-5119–94 (Knoxville, TN: Oak Ridge National Laboratory, 1994); Russell Lee, *Externalities and Electric Power: An Integrated Assessment Approach* (Oak Ridge, TN: Oak Ridge National Laboratory, 1995).

66. Gordon Walker, *Environmental Justice: Concepts, Evidence, and Politics* (London: Routledge, 2012).

67. Quoted in the Rock Ethics Institute at Pennsylvania State University, "Ethics in Climate Change," 2010, http://www.psu.edu/dept/rockethics/climate/.

68. Frank Biermann and Ingrid Boas, "Protecting Climate Refugees: The Case for a Global Protocol," *Environment* 50, no. 6 (2008): 8–16.

69. Joanna Kakissis, "Environmental Refugees Unable to Return Home," *The New York Times*, January 3, 2010, 23.

70. Kay Weir, "Don't Cry for Kiribati, Tuvalu, Marshall Islands, Parts of Papua New Guinea, the Caribbean, Bangladesh, Africa . . .," *Pacific Ecologist* (Winter 2008): 2.

71. Alex Smith, "Climate Refugees in Maldives Buy Land," Tree Hugger press release, November 16, 2008.

72. A. Rawlani and B. K. Sovacool, "Building Responsiveness to Climate Change through Community Based Adaptation in Bangladesh," *Mitigation and Adaptation Strategies for Global Change* 16, no. 8 (2011): 845–863.

73. Dale Jamieson, "Energy, Ethics, and the Transformation of Nature," in *The Ethics of Global Climate Change,* ed. Denis G. Arnold (Cambridge: Cambridge University Press, 2011), 16–37.

74. Per Australian Department of Foreign Affairs and Trade, "approximately 41 percent of the population lives on less than $1 a day," http://www.dfat.gov.au/geo/bangladesh/bangladesh_country_brief.html.

75. Ministry of Housing, Spatial Planning and the Environment; Ministry of Transport, Public Works and Water Management; Ministry of Agriculture, Nature and Food Quality; Ministry of Economic Affairs; National Programme for Spatial Adaptation to Climate Change, VROM 7222 (April 2007).

76. Sally Jenkins, "Sandy's Destruction in New York Turns Life in Lower Manhattan on Its Head," *The Washington Post*, October 30, 2012.

77. Mary Williams Walsh and Nelson Schwartz, "Estimate of Economic Losses Now Up to $50 Billion," *The New York Times*, November 1, 2012.

78. John Heilemann, "Sandy for the Suitors," *New York Magazine*, November 3, 2012.

7 The technological dimension

Efficiency, reliability, safety, and vulnerability

It is indeed true that we live in a society of risky choices, but it is one in which only some do the choosing, while others do the risking.

– Slavoj Zizek, *First As Tragedy, Then As Farce*
(New York: Verso Press, 2009), 13

Introduction

In June 1982, a Soviet natural gas pipeline caught fire and detonated in Siberia, causing an explosion so large that intelligence operatives monitoring early warning satellites in the United States thought that the blast was the detonation of a nuclear weapon. The cause of the explosion, it turned out, was a malfunction in the pipeline's computer control system. A few years earlier, Soviet intelligence operatives had apparently stolen pipeline supervisory control and data acquisition software from Canada, not knowing that U.S. Central Intelligence Agency (CIA) agents had tampered with the software and "planted" it in Canada for the Soviets to steal. The bugged software reset pump speeds and valve settings, producing pressures which eventually caused the pipeline to malfunction. The result has been described as the "most monumental non-nuclear explosion and fire ever seen from space."[1] The anecdote is apt for it reveals how energy infrastructure, national security, safety, and reliability are intertwined, so much so that spies explicitly targeted pipelines to weaken an enemy.

Only recently did people begin associating energy with images of roof-mounted solar panels, smart meters, and wind turbines. Before that, the energy sector was portrayed to the public eye – and, apparently, treated by the CIA in their sabotage efforts – as smokestacks, pipelines, and cooling towers. Despite the changes in the façade, the foundation and most of the load-bearing walls remain the same today as they were at the turn of the last century. The current energy system is highly centralized and structurally set up to capitalize on economies of scale. It depends on fossil fuels that for the most part rely on centralized transportation and distribution. And, as Slavoj Zizek's remark suggests, it segregates risk, with those that choose to invest in particular energy systems – bankers, policymakers, venture capitalists, engineers – very often different from those that must deal with the consequences of when those systems break-down, such as homeowners, minorities, and onsite company employees.

Although renewable and distributed technologies have seen steady if not remarkable growth, most energy demand is met by conventional technologies.[2] Rather than break with tradition, the developing world is trying to "catch up" with developed countries by duplicating such centralized systems.[3] The implications of this approach with regard to questions of justice and security are the central theme of this chapter. We start the chapter with examining the efficiency, reliability, safety, and vulnerability of energy systems in developing and developed countries. We include the threat of nuclear proliferation in our vulnerability discussion. The purpose of this inquiry is to determine the extent of the technological gap between the developed and developing worlds and assess whether bridging it with conventional technologies will provide an equitable solution. We then examine the reasons behind efficiency, reliability, safety, and vulnerability shortcomings and the justice implications of conventional energy systems. We end with the conclusion that blindly copying conventional energy systems places the world on an unjust and dangerous path dependence.

Energy technology in the developing world

For a vivid illustration of why maintaining the status quo in terms of energy development in poor countries is not an option, consider the case of Kathmandu, Nepal. Even though Thomas Edison began generating electricity more than 140 years ago, most of the 3 million residents of Kathmandu have never enjoyed reliable power. Figure 7.1 shows vendors at the night bazaar selling fish by candlelight in 2010. Nepal is not alone, however. As we

Figure 7.1 A butcher selling fish by candlelight in Kathmandu, Nepal, October 2010
Source: Benjamin K. Sovacool.

detailed in Chapter 4, billions of people in Asia, Africa, and South and Central America need access to electricity. In fact, in the uncertain world of energy forecasts, the projection that energy demand will grow predominately in non-OECD countries is probably the most certain thing.[4] Indeed, recent projections suggest that by the year 2030, more than 70 percent of all future greenhouse gas emissions will come from non-OECD countries.[5]

Efficiency

Overall, developing countries lag behind the developed world in terms of energy efficiency. They are less efficient using energy in industry (although the latest plant designs show significant efficiency gains), and though they hold their own in household use due to higher occupancy rates and more compact living, they also tend to have less efficient household appliances and buildings.[6] Regarding fossil fuel-based electricity production, developing countries are 6 percent less efficient than developed countries, with an average thermal efficiency of 33 percent.[7] However, the key to power generation efficiency is not technology but fuel mix. Countries with a larger share of natural gas tend to be more efficient than the ones relying on coal.[8] Russia serves as a rare but notable exception. Seventy-one percent of Russia's power generation comes from natural gas, but the country's overall power generation efficiency averages 33 percent. In contrast, China's efficiency in the electrical power sector is close to that of Russia's, with 97 percent of its electricity coming from coal.[9] Also, developing countries tend to lose 1 to 2 percent of GDP growth potential due to blackouts, over-investment in backup electricity generators, and inefficient use of resources.[10]

Reliability

Energy systems in developing countries usually lack reliability. Nigerians, for instance, live with such persistent power outages that one government official characterized the power supply as "epileptic."[11] The Nepal Electricity Authority, the state-owned monopoly supplier of power for the country, supplies electricity to the nation's capital Kathmandu for fewer than eight hours per day, with load shedding accounting for the remaining 16 hours.[12] When walking around the bazaars at night, more shops rely on candles to light their wares than electric bulbs, one of them depicted in Figure 7.1. In February 2010, the Philippines suffered a nationwide electricity breakdown that spanned three weeks of plant failures and rolling blackouts. The breakdown was caused by El Niño-induced droughts that curbed the country's hydroelectric capacity; it was exacerbated by human error, historical under-investment in the energy sector, improper facility maintenance, and delayed emergency response measures.[13]

Safety

Developing countries also face significant energy safety concerns. Many oil refineries, gas pipelines, power plants, hydroelectric dams, and coal mines are prone to unexpected breakdowns and, at times, catastrophic failure. To list a few prominent examples:[14]

- On September 29, 1957, in Kyshtym, Chelyabinsk, the former Soviet Union, heat exchangers failed at a liquid high-level nitrate storage tank at the Mayak Scientific-Production Association spent-fuel rod storage facility. This caused a chemical explosion that released 80 tons of radioactive waste, which then created a plume extending for 900 square kilometers. Authorities evacuated 270,000 people, and large amounts of agricultural cropland were contaminated. The accident caused 103 fatalities and $1.7 billion in damages.
- On August 8, 1975, in Henan Province, China, the Shimantan Dam failed during a severe storm and released 15,738 billion tons of water, causing widespread flooding that destroyed 18 villages and 1,500 homes, and induced disease epidemics and famine. The accident was responsible for 171,000 fatalities and caused $8.7 billion in damages.
- On August 11, 1979, in Gujarat State, India, the Machchu-2 Hydroelectric Dam situated on the Macchu River failed during a flood, inundating the industrial town of Morvi five kilometers downstream. The accident resulted in 1,500 fatalities, forced the evacuation of 150,000 people, and caused $2.4 billion in damages.
- On April 26, 1986, in Chernobyl, Ukraine, a mishandled reactor safety test caused a steam explosion and meltdown, necessitating the evacuation of 300,000 people from Kiev and dispersing radioactive material across Europe. The accident caused at least 4,056 fatalities and $7.2 billion in damages.
- A Nigerian National Petroleum Corporation high-pressure pipeline carrying gasoline ruptured and exploded in 1998, killing more than 1,000 people and causing $54 million in property damages.

The Paul Scherrer Institute (PSI) collected data on major industrial accidents from 1945 to 1996 and recorded 13,914 incidents.[15] The database included energy-related accidents that involved one of the following: at least five fatalities, at least 10 injuries, 200 evacuees, 10,000 tons of hydrocarbons released, more than 25 square kilometers of cleanup, or more than $5 million in economic losses. As listed in Table 7.1 9 out of 10 major energy accidents occurred in developing countries.[16]

These "major" and "headline-making" accidents obscure a set of more "chronic" or "everyday" accidents in developing countries. One of the best examples comes from China, where at least 45,000 coal miners perish every decade. In some individual years, such as 2002, more than 7,000 miners died – an average of 19 deaths *per day*, to say nothing of the hundreds of thousands of miners suffering from pneumoconiosis (black lung disease). As one study recently surmised:

> The insatiable demand for coal in China has not only led to well-established large-scale mines greatly exceeding safe production levels, it has also encouraged the growth of small-scale illegal mining by unscrupulous business people eager for profit and unconcerned with the lives of others.[17]

Table 7.1 Ten major energy accidents with the most immediate fatalities, 1969 to 1996

Energy system	Date	Location	Stage of energy chain	Immediate fatalities	Injuries	Evacuees	Cost (in millions 1996 $)
Oil	December 20, 1987	Philippines	Transport to refinery	3,000	26	0	–
Oil	November 1, 1982	Afghanistan	Regional distribution	2,700	400	0	–
Hydroelectric	August 11, 1979	India	Power plant	1,500	–	150,000	$1,024
Hydroelectric	August 27, 1993	China	Power plant	1,250	336	–	$27
Hydroelectric	September 18, 1980	India	Power plant	1,000	–	–	–
LPG	June 4, 1989	Russia	Long distance transport	600	755	0	–
Oil	November 2, 1994	Egypt	Regional distribution	580	–	–	$140
Oil	February 25, 1984	Brazil	Regional distribution	508	150	2,500	–
Oil	June 29, 1995	South Korea	Regional distribution	500	952	0	–
LPG	November 19, 1984	Mexico	Regional distribution	498	7,231	200,000	2.9

Source: Paul Scherrer Institute.

The causes behind these smaller-scale yet perpetual energy accidents include overproduction at regulated mines, poorly enforced safety standards, corruption, and poverty. The vast majority of miners in China today are rural migrants from poor areas and workers laid off from urban areas. They are paid according to how much coal they produce, have no social security or welfare benefits, and usually have to buy their own tools and protective equipment. The risks these miners face, and the casualties that result, will only grow in tandem with rising Chinese, and global, energy demand.

Vulnerability and nuclear proliferation

Though only a few developing countries have nuclear reactors, many more face the risk of nuclear weapons proliferation, mainly because the criminal and terrorist networks seeking such material often use developing countries as transit points to launch their operations.[18] From 1993 to 2003, in the aftermath of the collapse of the Soviet Union, authorities documented 917 incidents of nuclear smuggling in Russia, Germany, France, Turkey, Libya, Jordan, and Iran, with some of the more infamous cases presented in Table 7.2 (along with other, more recent examples).[19] Based on data available from the International Atomic Energy Agency (IAEA), India has reported 25 cases of missing radioactive materials. Of these, 13 have never been recovered and 52 percent have occurred by theft.[20] The Database on Nuclear Smuggling, Theft and Orphan Radiation Sources has reported that more than 40 kilograms of weapons-usable uranium and plutonium have been stolen from poorly protected nuclear facilities in the former Soviet Union

Table 7.2 Cases of nuclear theft and terrorist acquisition of fissile material, 1991 to 2012

Year	Location	Description
1991–1992	Glazov, Russia	Several kilograms of natural and depleted uranium are stolen from the Chepetsky Metallurgical Plant. The material is intercepted while being smuggled across the border to Poland. Twelve people, including plant personnel and security officers, are arrested.
1992	Podolsk, Russia	Leonid Smirnov, a chemical engineer, steals 1.5 kilograms of highly enriched uranium from the Luch Scientific Production Association over the course of five months and attempts to sell it to Middle Eastern suppliers before being caught by police.
1992	Khabarovsk Krai, Russia	Inspectors at the Komsomolska-na-Amure weapons depot discover that 23 nuclear warheads are unaccounted for.
1993	Kola Peninsula, Russia	Servicemen and poorly paid security guards conspire to steal two naval reactor fuel assemblies containing 3.6 kilograms of enriched uranium from the Andreeva Guba Technical Base.

(Continued)

Table 7.2 (Continued)

Year	Location	Description
1993	Izhevsk, Russia	Authorities apprehend a Lithuanian businessperson carrying 60 kilograms of weapons-grade uranium and a fuel rod in the trunk of his car, believed to be stolen from the Chepetsk Mechanical Plant one year earlier.
1993	Murmansk, Russia	Three naval reactor fuel assemblies containing 4.5 kilograms of enriched uranium are stolen from a fuel storage area at the Sevmorput Shipyard.
1995	Amman, Jordan	Agents from the U.S. Central Intelligence Agency seize 100 kilograms of enriched uranium, gyroscopes, and detonators from a shed at Amman Airport.
1995	Moscow, Russia	Chechen terrorists successfully smuggle a dirty bomb composed of C4 plastic explosive and Cesium-136 into a heavily populated Moscow park, but the bomb fails to detonate.
1998	Rome, Italy	Italian police seize 10 kilograms of highly enriched uranium and two fuel rods in the apartment of an African businessman believed to be stolen from a research reactor in the Congo.
1998	Istanbul, Turkey	Turkish detectives arrest six people for smuggling 13 glass tubes believed to contain nuclear material from Iran.
2001	Bogota, Columbia	Columbian police seize 0.5 kilograms of weapons-grade plutonium in the trunk of a car thought to be smuggled from Iran.
2001	Istanbul, Turkey	Turkish police size 2 kilograms of osmium-187 in 64 glass tubes (heat-resistant osmium can be combined with plutonium for coating the warheads of nuclear missiles).
2002	Moscow, Russia	Gen. Lebed announces that an inventory of Russia's Special Atomic Demolition Munitions, portable, low-yield nuclear weapons that are the equivalent of the United States' tactical nuclear weapons, revealed that 84 were missing.
2004	Prague, Czech Republic	Czech police seized 3 kilograms of highly enriched uranium from the apartment of a Syrian businessperson.
2004	Kiev, Ukraine	Ukrainian authorities seize 6 kilograms of U-235 from the apartment of two former Russian soldiers.
2005	Crimea, Ukraine	Ukrainian police confiscate six metal containers filled with cesium-137 inside a small village house.
2006	Tbilisi, Georgia	A Russian man and several Georgian accomplishes attempted to sell 100 grams of weapons grade uranium to an undercover CIA agent.
2011	Moldova, Moldova	Moldovan police arrest six people for trying to sell 1 kilogram of uranium-235 for $20 million.

Source: U.S. Government Accountability Office, U.S. Congressional Research Service, Benjamin K. Sovacool, and Marina Koren.

during the last decade. Similarly, the Nuclear Safety and Security Division of the IAEA manages an "Incident and Trafficking Database" (ITDB) that documents "incidents of illicit trafficking and other unauthorized activities and events involving nuclear and other radioactive material outside of regulatory control."[21] From January 1993 to December 2012, the ITDB recorded a sobering 2,331 incidents (with 160 incidents during the 2012 calendar year), of which 419 were confirmed to actually involve the criminal, unauthorized possession of nuclear materials. The ITDB classified 16 of these incidents as "serious" for involving the illegal trade of highly enriched uranium or plutonium. Indeed, it is telling that of the 71 countries currently engaged in nuclear activities, only 63 adhere to IAEA safeguards concerning the security of spent fuel and radioactive material.[22]

A discussion of nuclear proliferation segues naturally into a discussion of how vulnerable energy systems in developing countries are to "human-made" disruptions. Many developing countries suffer from political instability and high crime rates. Vulnerability of energy systems to theft becomes especially important when a single transmission line provides power to a neighborhood of tens and sometimes hundreds of thousands of people. As evidenced by the following examples, energy systems in developing countries display a high degree of vulnerability.

For instance, Chechen separatists financed their campaign against Russian federal forces by tapping the Baku-Grozny-Novorossiysk pipeline and diverting oil to clandestine refineries. The scope of the damage to the 480-mile Cano Limon-Covenas pipeline in Colombia rivals that of the Baku-Grozny-Novorossiysk pipeline. The pipeline has had so many holes in it that the local Colombians jokingly call it "the flute." Suicide bomber attacks on natural gas facilities continually recur in Nigeria, Sri Lanka, and Yemen. Infractions on energy infrastructure sometimes go beyond the actual equipment: in Pakistan, gunmen have frequently kidnapped employees of the Water and Power Development Authority and stormed Pakistan Petroleum Limited natural gas facilities.[23]

In developing nations, the impact of "human-made" infractions against energy facilities has been enormous, both in the scope of damage and the degree of disruption. Militants from the Farabundo Marti National Liberation Front were able to interrupt more than 90 percent of electric service in El Salvador. In the winter of 2002–2003, Venezuela's entire oil sector came to a screeching halt due to a strike at Petróleos. The effects radiated beyond the borders of the South American nation, as global oil supply went down by 3 percent. An attack on the Abqaiq oil processing facility[24] by al Qaeda terrorists in February 2006 slowed production for months, thereby pushing oil prices up $2 per barrel.[25] The Nigerian National Petroleum Corporation recorded more than 6,000 pipeline attacks from 2001 to 2007, and has a record-setting 3,000 pipeline attacks in 2011 – interruptions in energy supply that cost the government about $1 billion per *month* in lost revenue, no small sum when average per capita incomes are less than $2,700.[26] Most recently, in January 2013 Islamist extremists from Mali drove across the border in unmarked trucks and seized the internationally managed

Algerian *In Amenas* gas compression plant – the terrorists donned explosive-laden vests and used their AK-47 rifles to kill 37 employees of British Petroleum, Statoil, and the Algerian national oil company Sonatrach – suspending production for two weeks at a cost of at least US$100 million.[27] The siege ended only when the extremists threatened to explode the entire plant – which would have leveled a 25 square kilometer radius – prompting Algerian Special Forces to storm the complex, killing 29 militants

Clearly, energy systems in the developing world are, on average, inefficient, unreliable, unsafe, and vulnerable. Our next goal is to see whether energy systems in developed countries can provide a model and some inspiration for the developing world.

Energy technology in the developed world

Efficiency

By implementing efficiency measures, developing countries can certainly offset some impacts of growth in energy demand. However, efficiency gains in household energy consumption in the developed world are often negated by increased square footage per person, greater "plug loads" from televisions, air conditioners, and electric appliances, and decreased occupancy rates.[28] The same applies to the transportation sector where improvements in energy efficiency are offset by vehicle weight and traffic congestion. Also, as noted above, developed countries rely more on fossil fuels in the household energy fuel mix.[29] And while developed countries are generally far more efficient in industrial use – Japan and South Korea lead the charge, followed by Europe and North America – they are still shockingly inefficient.[30]

For example, one study found that converting coal at a power plant into an incandescent bulb inside a home wastes nearly 97 percent of the original energy potential.[31] Converting to a compact fluorescent bulb would improve this efficiency about fourfold, but would do little to account for the remaining inefficiencies. Similarly, a typical internal combustion engine in an automobile only utilizes about 10 percent of the energy content of the gasoline to move the car, with the rest of the energy, again, being lost as heat.[32] The nonpartisan U.S. Environmental Protection Agency estimated in 2011 that as much as one-third of all energy used in the commercial sector of the United States is "wasted or inefficiently used" and another study concluded that only 13 percent of primary energy currently used in the United States ends up as "useful energy services."[33] Additionally, about 20 huge power plants operate to energize U.S. appliances and equipment that are turned off, just to keep them in standby mode.[34]

Reliability

Though developed countries have more reliable energy systems generally, the costs when they breakdown are usually more substantial. With a more advanced economy, the cost of even a single short blackout (similar to the likes of the Northeast blackout of 2003 that affected 50 million Americans and Canadians) increase astronomically.[35] Between 1964 and 2005, for instance, the Institute of Electrical and Electronics Engineers estimated that no less than 17 major blackouts affected more than 195 million residential, commercial, and industrial customers in the United States, with seven of these major blackouts occurring in the final 10 years. Sixty-six smaller blackouts (affecting between 50,000 and 600,000 customers) occurred from 1991 to 1995, and 76 occurred from 1996 to 2000, and that is for the United States alone. The costs of these blackouts are monumental: The DOE estimates that power outages and power quality disturbances cost customers as much as $206 billion annually or more than the entire nation's electricity bill for 1990. Indeed, the U.S. transmission and distribution network has become so prone to failure that the American Society of Civil Engineers gave it a grade of "D+" and warned that it was in "urgent need of modernization."[36]

It is important to note that the frequency of outages determines their costs rather than their duration. In the modern digital economy, even short power outages can shut down communication devices, refrigerators, water and process pumping, the use of credit card and fax machines, and electronic cash registers. One chief executive from a global microchip company commented that:

> My local utility tells me they only had 20 minutes of outages all year. I remind them that these four five-minute episodes interrupted my process, shut down and burnt out some of my controls, idled my work force. I had to call my control service firm, call in my computer repair firm, direct my employees to test the system. They cost me eight days and millions of dollars.

The International Energy Agency estimated that the average cost of a one-hour power outage was $7.5 million for brokerage operations and $3.1 million for credit card companies. A 2005 survey found that one minute of downtime cost automotive businesses an average of $24,000. The figures grow dramatically for the semiconductor industry, where a two-hour power outage can cost close to $54 million.

The economic cost of the aforementioned Northeast U.S. blackout precipitated on August 14, 2003, came to approximately $5 per foregone kWh, a figure that was roughly 50 times greater than the average retail cost throughout the country at that time. Initially 60,000 to 65,000 MW of load was interrupted (approximately 11 percent of the Eastern Interconnection), and eventually 531 generators were shut down at 263 plants. Economists have estimated the total loss at more than $6 billion, with $1 billion of losses spread across New York City alone.[37] The root cause analysis of the blackout cited several deficiencies: inadequate "situational awareness" at First Energy Corporation in Indiana, where the blackout began; failure to adequately trim trees in First Energy's transmission right of way; and failure of reliability coordinators to promptly identify and deal with problems.

Safety

Energy systems in the developed countries may be *safer* than in developing countries but they are also accident-prone. And, as evidenced here, these accidents cause catastrophic damage and come at astronomical cost.

• On March 24, 1989, in Alaska, United States, the Exxon oil tanker *Valdez* ran aground, its captain Joseph Jeffrey Hazelwood allegedly under the influence of alcohol, and spilled 250,000 barrels into Prince William Sound. The resulting oil spill spread as far as 450 miles (724 km) away from Prince William Sound, killed 250,000 seabirds, 4,000 sea otters, 250 bald eagles, and more than 20 orca whales. Herring fishery losses alone amount to more than $286 million, and $4.7 billion was spent on cleanup costs and fines. Twenty years later, some locations remain as toxic as they were in 1989. An extensive study conducted by NOAA in 2001 dug more than 9,000 shallow holes at 91 sites over a period of 3 months, and confirmed that oil turned up in half of the pits.[38]

• On November 13, 2002, near Galicia, Spain, the oil tanker *Prestige* ruptured one of its tanks during severe weather and spilled 20 million gallons of oil into the sea and along French, Spanish, and Portuguese beaches. No fatalities occurred but authorities reported $3.7 billion in damages and remediation costs.

• On December 22, 2008, an earthen dike broke at a 40-acre waste retention pond at the Tennessee Valley Authority's Kingston Fossil Plant in Roane County, Tennessee, United States. More than 1 billion gallons of toxic coal fly ash slurry was released. The slurry (a mixture of fly ash and water) traveled downhill, covering 400 acres of the surrounding land, damaging homes, and flowing into nearby waterways such as the Emory River and Clinch River (tributaries of the Tennessee River). It was the largest fly ash slurry spill in U.S. history, and involved property damages and cleanup costs exceeding $975 million (see Figure 7.2).

• On April 20, 2010 in the Tiber Oil Field, Gulf of Mexico, United States, an explosion on board British Petroleum's *Deepwater Horizon,* the deepest offshore oil well in the world, caused a fire that burned for 36 hours before sinking the entire rig. Oil continued to spill from the well in massive quantities until engineers sealed the well on September 19, 2010. Eleven crewmembers died in the explosion, and damages, litigation, and cleanup costs totaled approximately $41 billion.

• On March 11, 2011, an earthquake and tsunami in Fukushima Prefecture, Japan caused emergency backup generators to malfunction at the Fukushima Daiichi nuclear power plant. The pressure vessels at some of its reactors cracked and, as a result, spent fuel pools at the facility caught fire, fuel assemblies melted down, and dangerous radioactive releases occurred. Twenty-one fatalities occurred among first responders and evacuees (with more than 200,000 evacuated), along with $152 billion in damages.

Another study, published in the May 2008 issue of *Energy Policy* (and written by one of the authors of this book), assessed major energy accidents worldwide from 1907 to 2007.[39] The study identified 279 incidents totaling $41 billion in

Figure 7.2 An aerial view of the kingston fly ash spill, December 2008
Source: Tennessee Valley Authority.

damages and 182,156 fatalities, with the number of accidents peaking in the decade between 1978 and 1987, which had more than 90 accidents.[40] The study found that out of 30 major accidents having more than $100 million in damages, 25 occurred in developed countries (Table 7.3). The study also noted that 18 out of 30 accidents that caused more than 100 fatalities took place in developed countries (Table 7.4). Even more telling, the study ended in 2007, meaning it did

Table 7.3 Major energy accidents with more than $100 million in damages, 1907 to 2007

Facility	Date	Location	Description	Cost (in millions 2006 $)
Coal mine	December 6, 1907	Monongah, West Virginia, United States	Underground explosion traps workers and destroys railroad bridges leading into the mine	$162
LNG Plant	October 20, 1944	Cleveland, Ohio, United States	Explosion at LNG Facility destroys one square mile of Cleveland	$890
Oil tanker	June 13, 1968	Economic Exclusive Zone, South Africa	The World Glory oil tanker experiences full failure, spilling 11 million gallons of fuel	$110
Hydro-electric	August 8, 1975	Henan Province, China	Shimantan Dam fails and releases 15,738 billion tons of water, causing widespread flooding that destroys 18 villages, 1,500 homes, and induces disease epidemics and famine	$8,700
Nuclear	November 24, 1989	Greifswald, East Germany	Electrical error causes fire in the main trough that destroys control lines and five main coolant pumps, and almost induces meltdown	$443
Hydro-electric	June 5, 1976	Idaho Falls, Idaho, United States	Teton dam fails and releases 300,000 acre feet of water that floods farmland and towns surrounding Idaho Falls	$990
Oil tanker	December 15, 1976	Nantucket, Massachusetts, United States	The oil tanker Argo Merchant runs aground, spilling 7.5 million gallons of fuel and causing an oil slick 100 miles long and 70 miles wide	$267
Nuclear	February 22, 1977	Jaslovske Bohunice, Czechoslovakia	Mechanical failure during fuel loading causes severe corrosion of reactor and release of radioactivity into the plant area, necessitating total decommission	$1,700
Oil tanker	March 16, 1978	Porstall, Brittany, France	The oil tanker Amoco Cadiz experiences steering failure and runs aground, spilling 68 million gallons of fuel	$111
Oil tanker	July 19, 1979	Tobago, Caribbean	Two Very Large Crude Carriers, the Atlantic Empress and Aegean Captain, collide at sea, spilling 111 million gallons of fuel	$120
Nuclear	March 28, 1979	Middletown, Pennsylvania, United States	Equipment failures and operator error contribute to loss of coolant and partial core meltdown at Three Mile Island nuclear reactor	$2,400

Table 7.3 (Continued)

Type	Date	Location	Description	
Oil tanker	August 6, 1983	Cape Town, South Africa	The oil tanker *Castillo de Bellver* catches fire, spilling 75 million gallons	$143
Nuclear	September 15, 1984	Athens, Alabama, United States	Safety violations, operator error, and design problems force six-year outage at Browns Ferry Unit 2	$110
Nuclear	March 9, 1985	Athens, Alabama, United States	Instrumentation systems malfunction during startup, convincing the Tennessee Valley Authority to suspend operations at all three Browns Ferry Units	$1,830
Nuclear	December 26, 1985	Clay Station, California, United States	Safety and control systems unexpectedly fail at Rancho Seco nuclear reactor, ultimately leading to the premature closure of the plant	$672
Nuclear	April 11, 1986	Plymouth, Massachusetts, United States	Recurring equipment problems with instrumentation, vacuum breakers, instrument air system, and main transformer force emergency shutdown of Boston Edison's Pilgrim nuclear facility	$1,001
Nuclear	April 26, 1986	Kiev, Ukraine	Mishandled reactor safety test at Chernobyl nuclear reactor causes steam explosion and meltdown, necessitating the evacuation of 300,000 people from Kiev and dispersing radioactive material across Europe	$6,700
Nuclear	May 4, 1986	Hamm-Uentrop, Germany	Operator actions to dislodge damaged fuel rod at Experimental High Temperature Gas Reactor release excessive radiation to 4 square kilometers surrounding the facility	$267
Nuclear	March 31, 1987	Delta, Pennsylvania, United States	Philadelphia Electric Company shuts down Peach Bottom Units 2 and 3 due to cooling malfunctions and unexplained equipment problems	$400
Nuclear	December 19, 1987	Lycoming, New York, United States	Fuel rod, waste storage, and water pumping malfunctions force Niagara Mohawk Power Corporation to shut down Nine Mile Point Unit 1	$150
Oil platform	July 6, 1988	North Sea, United Kingdom	A natural gas leak causes explosion on Occidental Petroleum's *Piper Alpha* oil rig	$190
Nuclear	March 17, 1989	Lusby, Maryland, United States	Inspections at Baltimore Gas & Electric's Calvert Cliff Units 1 and 2 reveal cracks at pressurized heater sleeves, forcing extended shutdowns	$120

(*Continued*)

Table 7.3 (Continued)

Facility	Date	Location	Description	Cost (in millions 2006 $)
Oil tanker	March 24, 1989	Prince William Sound, Alaska, United States	The Exxon oil tanker *Valdez* runs aground and spills 250,000 barrels into Prince William Sound	$4,100
Oil tanker	June 5, 1989	Canary Islands, Spain	The oil tanker *Kharg 5* catches fire and spills 19,500,000 gallons of oil	$165
Oil tanker	January 21, 1993	Andamen Sea	The oil tanker *Maersk Navigator* collides with the *Sanko Honour*, spilling 2 million barrels of oil	$169
Oil tanker	February 15, 1996	Pembroke-shire, Wales, United Kingdom	The oil tanker *Sea Empress* runs aground and spills 19 million gallons of fuel	$114
Nuclear	February 20, 1996	Waterford, Connecticut, United States	Leaking valve forces Northeast Utilities Company to shut down Millstone Units 1 and 2; further inspection reveals multiple equipment failures	$254
Nuclear	September 2, 1996	Crystal River, Florida, United States	Balance-of-plant equipment malfunction forces Florida Power Corporation to shut down Crystal River Unit 3 and make extensive repairs	$384
Nuclear	February 16, 2002	Oak Harbor, Ohio, United States	Severe corrosion of control rod forces 24-month outage of Davis-Besse reactor	$143
Oil tanker	November 13, 2002	Galicia, Spain	The oil tanker *Prestige* ruptures one of its tanks during severe weather and spills 20 million gallons of oil into the sea and along French, Spanish, and Portuguese beaches	$3,300

Source: Benjamin K. Sovacool.

Table 7.4 Major energy accidents with more than 100 fatalities, 1907 to 2007

Facility	Date	Location	Description	Fatalities
Coal mine	December 6, 1907	Monongah, West Virginia, United States	Underground explosion traps workers and destroys railroad bridges leading into the mine	362
Coal mine	February 27, 1908	San Juan de Sabinas, Coahuila, Mexico	Mine shaft completely collapses	201
Coal mine	September 30, 1908	Palau Coal Mine, Coahuila, Mexico	Explosion and fire collapse multiple shafts	100
Coal mine	February 16, 1909	Stanley, England	Explosion and fire destroys entire mine	168
Coal mine	November 13, 1909	Cherry, Illinois, United States	Fire and explosion collapse multiple shafts	259
Coal mine	October 22, 1913	Dawson, New Mexico, United States	Fire induces explosion that buries workers	263
Coal mine	October 14, 1913	Cardiff, Wales, England	Mine shaft completely collapses	439
Coal mine	June 19, 1914	Hillcrest, Alberta, Canada	Fire and explosion collapse multiple shafts	189
Hydroelectric	December 1, 1923	Valle di Scalve, Italy	Gleno's Dam complete fails, flooding the local countryside	202
Coal mine	March 8, 1924	Castle Gate, Utah, United States	Three explosions destroy entire mine	172
Coal mine	September 22, 1934	Gresford, Wrexham, Wales, England	Fire and explosion destroy entire mine	266
Natural gas pipeline	March 18, 1937	New London, Texas, United States	Natural gas explosion destroys entire high school	309
LNG Plant	October 20, 1944	Cleveland, Ohio, United States	Explosion at LNG Facility destroys one square mile of Cleveland	130
Coal mine	June 19, 1945	Rancagua, Chile	Smoke from fire suffocates miners	355
Coal mine	March 25, 1947	Centralia, Illinois, United States	Fire and explosion destroy entire mine	111
Coal mine	December 21, 1951	West Frankfort, Illinois, United States	Main shaft caves in, trapping workers	119
Coal mine	August 8, 1956	Marcinelle, Belgium	Fire and explosion destroy facility	262
Coal mine	February 7, 1962	Volkingen, Germany	Simultaneous methane and coal dust explosion destroys half of the facility	299
Coal mine	March 31, 1969	Coahuila, Mexico	Flooding causes avalanche at Mina de Barroterán coal mine	176

(Continued)

Table 7.4 (Continued)

Facility	Date	Location	Description	Fatalities
Coal mine	February 26, 1972	Logan County, West Virginia, United States	Coal slurry impoundment damn fails and releases 132 million gallons of black wastewater into the surrounding communities	125
Hydroelectric	August 8, 1975	Henan Province, China	Shimantan Dam fails and releases 15,738 billion tons of water causing widespread flooding that destroys 18 villages and 1,500 homes, and induces disease epidemics and famine	171,000
Oil platform	March 27, 1980	Ekofisk oil field, North Sea, United Kingdom	The oil rig *Alexander Keilland* breaks apart under fatigue and capsizes, killing entire crew	123
Nuclear	April 26, 1986	Kiev, Ukraine	Mishandled reactor safety test at Chernobyl nuclear reactor causes steam explosion and meltdown, necessitating the evacuation of 300,000 people from Kiev and dispersing radioactive material across Europe	4,056
Oil platform	July 6, 1988	North Sea, United Kingdom	A natural gas leak causes explosion on Occidental Petroleum's *Piper Alpha* oil rig	167
Oil pipeline	June 4, 1989	Ufa, Russia	Sparks from passing trains ignite gas leaking from petroleum pipeline, causing multiple explosions that derail both trains	643
Oil pipeline	October 17, 1998	Niger Delta, Nigeria	Petroleum pipeline ruptures and explodes, destroying two villages and hundreds of villagers scavenging gasoline	1,078
Coal mine	February 14, 2005	Fuxin, China	Gas explosion destroys Liaoning coal mine, trapping workers inside	210
Oil pipeline	May 12, 2006	Lagos, Nigeria	An oil pipeline ruptures and spills dirty diesel fuel, causing a fire that destroys three villages	143
Oil pipeline	December 26, 2006	Lagos, Nigeria	Oil pipeline explodes, causing widespread fires that destroy more than 300 homes	466
Coal mine	March 19 2007	Novokuznetsk, Russia	Methane gas causes explosion at Ilyanovskaya mine	110

Source: Benjamin K. Sovacool.

not include at least two major energy accidents – the *Deepwater Horizon* oil spill in the Gulf of Mexico and the Fukushima nuclear power meltdown in Japan, which by *themselves* have each caused more than $41 billion in damages.

Vulnerability and nuclear proliferation

Despite a popular opinion fostered by Hollywood films, nuclear proliferation concerns do not end at old Soviet nuclear facilities. The truth is that nuclear facilities in developed countries display a high degree of vulnerability. In the United States, the nuclear industry has been criticized for the weak allegiance of its workers, the high turnover rate (most work at their job for less than six months), poor background checks, inferior equipment, poor communication with local authorities, and little training.[41] One nuclear facility hired a convicted felon who had been imprisoned for armed robbery.[42] Worker sabotage has been reported at no less than eight American reactors: Indian Point (New York), Zion (Illinois), Quad Cities (Illinois), Peach Bottom (Pennsylvania), Fort St. Vrain (Colorado), Trojan (Oregon), Browns Ferry (Alabama), and Beaver Valley (Pennsylvania).[43] Nuclear operators have been found falsifying reactor records to escape fines and cheating on licensing examinations (including two of the operators that caused the accident at Three Mile Island). As nuclear security officer Richard Kester quipped, "Charles Manson could get access to a nuclear power plant."[44]

Yet, despite the history of past incidents, attempts to curb safety shortcomings have yet to succeed. Industry-wide exercises in the United States in 2002 found that 37 of 81 nuclear plants included in the program did not meet their Operational Safeguards Readiness Evaluation.[45] Things have yet to improve on the other side of the Atlantic. In the United Kingdom, per the Office of Civil Nuclear Security, Greenpeace activists were able to walk into the Sizewell B nuclear power plant site in Suffolk to highlight the facility's lack of security. However, even after the two initial incursions, the organization's activists were able to repeat their successful trips to the plant's premises the following year.[46] In 2004 alone, more than 40 security breaches occurred at civil nuclear facilities throughout the United Kingdom.[47]

The September 11, 2001 terrorist attacks in the United States give good reason to question the ability of many existing nuclear plants to maintain integrity in case of an aircraft impact. Simply put, nuclear reactors were not designed to withstand the impact of a large modern aircraft loaded with fuel. According to a study by the Swiss Federal Nuclear Safety Inspectorate, the safest plants in Sweden were built to endure the impact of a Boeing 707 with a residual fuel level and an impact velocity of 370 kilometers an hour.[48] Thus, even the best designs are not capable of stopping a mammoth Boeing 747–8 or Airbus A-380 heading toward a reactor building at a speed of over 600 kilometers an hour.[49]

Vulnerability of energy systems in developed countries goes well beyond just nuclear facilities. In 1974, two extortionists in Oregon, United States threatened to cut power to Portland if they were not given $1 million. They backed up their threat by destroying 14 transmission towers in rural Oregon. Transmission tower bombings (unrelated to the Oregon incident) were also carried out in Alabama, California, Louisiana, New Jersey, Ohio, Washington, and Wisconsin.[50] According to the U.S. Federal Bureau of Investigation, 2 percent of the 15,000 bombings occurring each year in the United States are directed at electric utilities. The increase in bombings in 1978 was linked to overall frustration over higher energy prices. That year, an American utility was bombed every 12 days.[51]

Members of the Irish Republican Army cut transmission lines to aid jailbreak in Ireland. Militants from the same organization also plotted to disable gas pipelines and other utilities across London with 36 explosive devices. Fortunately for the rest of the inhabitants of Europe's largest city, Scotland Yard foiled the plot.[52] As in developing countries, oil and gas pipelines pose an especially attractive target for attacks. Members of the Ku Klux Klan have been convicted of attempting to attack natural gas infrastructure throughout the United States. The Pacific Gas and Electric Company's pipelines in California, United States, were bombed by the New World Liberation Front 10 times in 1975 alone. Vancouver police in Canada prevented a plot to assault the trans-Alaskan pipeline for personal profit in oil futures in 1999.[53] However, law enforcement agencies were not successful protecting the pipeline two years later. A man, who was later described in court as "a career criminal," fired at the pipeline with a high-powered rifle.[54] The bullet pierced through thick steel and several inches of high-density insulation, causing oil to stream out with great force.[55] The incident, illustrated with Figure 7.3, resulted in a two-day shutdown of

Figure 7.3 The site of the Trans-Alaskan pipeline rupture, 2001

Source: The U.S. Federal Bureau of Investigation.

the trans-Alaskan pipeline. More importantly, "[m]ore than 285,000 gallons of crude were spilled as a result of that small bullet hole and – according to press reports at the time – the cleanup took many months and cost $13 million."[56]

Energy systems in developed countries are generally more efficient, reliable, safe, and less vulnerable than those in developing countries. Developed countries get more out of their power plants; regular protracted blackouts and brownouts do not happen that often; fatalities rarely exceed a few hundred (as opposed to thousands); and oil and gas pipelines do not resemble flutes. However, this is hardly a reason for optimism and certainly no cause for celebration. Conventional energy systems in developed countries display the same basic flaw: they do not power economies and serve societies in the most efficient, reliable, safe, and secure way possible.

From structural problems to path dependence

The foregoing discussion intended to determine whether duplicating conventional technologies in developing countries will satisfy growth in energy demand in a more efficient, reliable, safe, and less vulnerable manner. The evidence suggests that it will not. In this section we look at conventional energy technologies in terms of the limits of efficiency, the reasons behind the lack of reliability and safety, the circumstances that increase vulnerability, and finally, the energy justice implications of all these factors.

Efficiency

When considering solutions for energy-related problems, it is sometimes remarked that the standard approach should be, "The answer is energy efficiency – what is the question?" Humor aside, energy efficiency is undoubtedly a critical and necessary part of the solution to the world's energy problems. Getting more kilometers per liter of gasoline, generating more BTUs per ton of coal, and losing fewer kWh per mile of a transmission line are the kind of options that should be carefully examined before considering other technological options. According to the IEA, "[e]nergy efficiency is widely recognized as a key option in the hands of policy makers."[57] Major energy-consuming nations from the developed and developing world have demonstrated a willingness to realize the potential that energy efficiency presents. For example, China is aiming to reduce its energy intensity by 16 percent by 2015.[58] The European Union has committed to becoming 20 percent more energy efficient by 2020.[59] The U.S. Energy Information Administration went as far as developing the Efficient World Scenario for the 2012 World Energy Outlook "to quantify the implications for the economy, the environment and energy security of a major step change in energy efficiency."[60]

However, relying solely on improving the energy efficiency of conventional systems has two major obstacles in terms of energy justice. First, to make something energy efficient, one must have something that is energy inefficient, i.e., have capacity for energy efficiency. There is no doubt that the livelihoods of 9 percent of Uganda's population would improve with appliances that need less electricity and transmission equipment that can minimize strain on transmission lines.[61] However, the economic and social gap between the lucky 9 percent and the rest of Ugandans would grow even wider, as the 91 percent lack access to electricity to begin with and cannot benefit from improvements in energy efficiency.

Secondly, although the current global capacity for energy efficiency is sizable, it is not infinite.[62] Therefore, deployment of energy efficiency alone will not prevent the world's poorest from getting hit with the catastrophic effects of global climate change.[63] According to the Energy Information Administration, "[i]f action to reduce CO_2 emissions is not taken before 2017, all the allowable CO_2 emissions would be locked-in by energy infrastructure existing at that time."[64] However, prompt implementation of energy efficiency measures in line with the pace, scale, and scope suggested by the Efficient World Scenario may delay this complete lock-in until 2022.[65]

Reliability

Our inquiry into the reliability of energy systems in the preceding two sections revealed matching patterns showing the propensity of energy infrastructure to service interruptions. The similarity originates from the centralized approach used to design energy systems. Moreover, the same approach is often applied to increase reliability of the infrastructure. Such measures lead to higher costs of service and an even more inequitable allocation of financial burdens. In addition, a "top-down" approach often precludes other means of improving reliability such as distributed generation.

As stated numerous times throughout the book, conventional energy systems are centralized, meaning that power plants and refineries are often far from users, which necessitates longer transmission and distribution distances. Electricity suppliers, for example, have clustered generating units geographically near oil fields, coal mines, sources of water, demand centers, and each other, interconnecting them sparsely and thereby making them heavily dependent on a few critical nodes and links. Operators have provided little storage to buffer successive stages of energy conversion and distribution, meaning that failures tend to be abrupt and unexpected rather than gradual and predictable. Firms have also located generators remotely from users, so that supply-chain links have to be long and the overall system lacks the qualities of user controllability, comprehensibility, and interdependence.[66] Scott Sklar, the president of an energy-consulting firm, used the following analogy to describe the existing heavily strained power grid:

> The electric utility sector is like the Space Shuttle. When it works, it works fantastically well. But when it fails, it screws up devastatingly. We

spend billions of dollars on surge protectors, which means our electric utility system doesn't have the quality it needs. Instead, issues of quality (and when it fails) are shoved onto the consumer. And it may only get worse: twenty years ago the country wasn't moving towards a digitized economy. Now our manufacturing controls, computers, and telecommunications devices (even our appliances) are starting to move towards complete digitization, which will place even more stress on the existing power grid.[67]

It is worth noting that computerization does not necessarily make the grid less reliable. Certain computerized smart grid technologies used in combination with distributed generation and storage can contribute to peak load shedding and better load management, actually making the system more reliable. However, automation and computerization with the goal of improving operational efficiency may also produce the opposite results. Low margins and various competitive priorities have encouraged industry consolidation, with fewer and bigger facilities and intensive use of assets centralized in one geographical area. As the National Research Council noted, "[power control systems are] more centralized, spare parts inventories have been reduced, and subsystems are highly integrated across the entire business."[68] Restructuring and consolidation have resulted in lower investment in reliability in recent years, as cash-strapped utilities seek to minimize costs and maximize revenue available for other areas.

Other energy systems are not well adapted to changes in throughput. Natural gas is a good example: if pressure falls below a certain level, a pipeline no longer works. Many forms of transport (pipelines, tankers) are built to move only one type of fuel. Also, many processing facilities are designed to handle only a certain type of raw material. For example, the main reason why Enbridge chose to push for extension of the Keystone XL pipeline in the United States is the available refining capacity for heavy crude in the Gulf of Mexico refineries.[69] When military conflict erupted in Libya in 2011 and the country lost 22 percent of its 1.65 million-barrel-a-day output, oil prices immediately hit a two-and-a-half-year high.[70] The ensuing near-total disruption of production provoked the United States to release 60 million barrels of oil from the emergency supply of its members, its first emergency release since Hurricane Katrina.[71] Such drastic consequences might seem puzzling given the world's daily oil output of 87.4 million barrels.[72] However, Libya produces mostly light, sweet crude that is processed by certain refineries into certain types of petroleum products. Thus, the shortfall of Libyan crude could not have been offset by Saudi Arabia's or Russia's giant fields.

This complexity and interdependence can be a curse as much as a blessing, for systems can interact in unforeseen and unplanned ways. Some technologies, dependent on multiple fuels at once, can fail due to lack of any one of them, such as a furnace that burns oil or natural gas but needs electricity to ignite and pump the fuel, or gasoline pumps that rely on electricity. When built too closely together, failures in one system can cascade to the other. Broken water mains, for instance, can short out circuits or electric cables; fires and explosions can ignite entire pipeline networks; earthquakes can cause gas mains to rupture and explode, destroying facilities that survived the initial shock. In addition to horrific injuries and property damage, these sorts of incidents lead to blackouts and brownouts and interruptions in supply.

The structural features of conventional centralized energy systems implicate questions of energy justice in three ways in relation to reliability. First, there is a built-in bias against distributed generation. The existing electric power system is designed to accommodate centralized dispatchable sources, often precluding proliferation of new, decentralized, low-carbon technologies. A report by the Intergovernmental Panel on Climate Change noted, "[T]here are few, if any, technical limits to the planned system integration of [renewable energy] technologies across the very broad range of present energy supply systems worldwide."[73]

Secondly, there is the manner in which reliability costs are allocated. Centralized supply for electricity cannot discriminate well between end-users. Electricity for a water heater is unaffected by a few hours of interruption, but must bear the high cost of extreme reliability needed for subways and hospitals. Consumers still pay to have the network operate as well as it does. Transmission and distribution (T&D) often accounts for more than twice the cost of generation on most consumers' electricity bills – underscoring how far electricity must travel from a typical power plant to its point of use. During hot weather, when power lines stretch and conductivity decreases, T&D losses can exceed 25 percent, and in countries such as Indonesia or Myanmar, losses can exceed 50 percent. The same reliability concerns occur in the provision of natural gas, which can be a bane for high priority customers such as hospitals.[74]

Finally, there is the disproportionate ability of different socioeconomic groups to offset reliability problems. Simply put, wealthier people are in a better position to buy electric generators and to pay more for fuel in times of service interruption or supply shortage. As we have seen many times in this book, it is the poor that get hit the worst by the deficiencies of the current energy system.

Safety

Although reliability concerns are serious and should not be underestimated, accidents raise questions regarding the fundamental safety of the current energy system. Often, accidents affect people located many miles from the site who in no way benefit from the faulty facility. Accidents that are more local in scale tend strike areas located nearby the troubled infrastructure, affecting communities with a high percentage of indigent population.[75]

Unfortunately, accidental failures of conventional energy infrastructure are an uncomfortable norm within the industry. The aforementioned PSI study concluded that an astounding 31 percent of major industrial accidents were related to the energy sector – the highest of any individual sector.[76] The

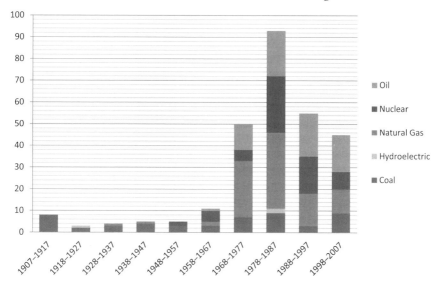

Figure 7.4 Major energy accidents by decade and source, 1907 to 2007
Source: Benjamin K. Sovacool.

authors of the study noted that the frequency and severity of accidents were related to a number of factors, including flaws in design, weak regulatory frameworks, poor quality assurance requirements, the absence of a culture of safety, and the degree of effectiveness of medical response. According to the previously mentioned study in *Energy Policy*, accidents at dams were the most dangerous, accidents at nuclear power plants the most expensive, and accidents at oil and gas pipelines the most frequent, as Figure 7.4 shows.[77]

As depicted in the figure above, oil, natural gas, and nuclear facilities are clear "leaders" in safety infractions. Oil and gas pipelines are prone to catastrophic mechanical failure from within and impact failure (things crashing into pipelines) from without. Faulty joints connecting pipeline components, malfunctioning valves, operator error, and corrosion lead to frequent leaks and ruptures. Examining data from 1907 to 2007, the *Energy Policy* study found that natural gas pipelines are the type of energy infrastructure most frequent to fail, accounting for 33 percent of all major energy accidents worldwide. Internationally, between 20 million and 430 million gallons of oil were spilled in reported incidents each year between 1978 and 1997, with the number of significant spills ranging from 136 to 382 annually over this period. The U.S. Department of Transportation has noted that oil and gas pipelines fail so often in the United States that they expect 2,241 major accidents and an additional 16,000 spills every 10 years. The PSI study, similarly, noted that the "riskiest" stages for oil appeared to be when it was being distributed through regional pipelines and trucks or transported to refineries.

These two stages accounted for more than three-fourths of all oil-related accidents. It also noted that nearly three-quarters of all natural gas accidents were associated with pipelines, and about 21 percent of these accidents involved mechanical failure.

Nuclear power is especially accident-prone. Because nuclear power plants are so large and complex, accidents onsite tend to be very expensive. The *Energy Policy* study found that 63 nuclear accidents have occurred worldwide from 1947 to 2007. The study documented that nuclear plants ranked highest in economic cost among all energy accidents, accounting for 41 percent of energy-accident-related property damage from 1907 to 2007 (or $16.6 billion). These numbers translate to more than one incident and $332 million in damages every year for the past three decades. Twenty-nine accidents have occurred since the Chernobyl disaster in 1986, and 71 percent of all nuclear accidents (45 out of 63) occurred in the United States, refuting the notion that severe accidents cannot happen within a highly regulated country or that they have not happened since Chernobyl. Such accidents have involved meltdowns, explosions, fires, and loss of coolant, and have occurred during both normal operation and extreme emergency conditions (such as droughts and earthquakes).

Unfortunately, nuclear accidents are not a thing of the past. According to former nuclear engineer David Lochbaum, the majority of serious nuclear accidents occurred with recent technology, making newer systems less safe.[78] This can be explained by the fact that many operators are less familiar with new technological solutions, which increases the risk of an accident. The future may not see much improvement in the safety of nuclear facilities, either. An interdisciplinary team at MIT conducted a study on possible reactor failures and concluded that, given the current rate of growth of nuclear power, at a minimum, four serious core damage accidents will occur from 2005 to 2055. The study concluded that "both the historical and probabilistic risk assessment data show an unacceptable accident frequency. The potential impact on the public from safety or waste management failure . . . make it impossible today to make a credible case for the immediate expanded use of nuclear power."[79]

Although coal-related accidents have not been among the top three accident types in the last three decades, they remain a prominent feature in global accident statistics. Notwithstanding improved safety regulations and medical treatment, underground coal mining is still a very dangerous industry. Miners frequently encounter pockets of underground methane gas, which is highly explosive and produced in significant quantities when coal is removed. Shifting and unpredictable geologic conditions often make the roofs of mines unstable, and miners always face the ever-present risk of flooding and fire, hazards that have increased in recent years as miners dig deeper to reach coal reserves.[80]

There is little doubt that the technology behind conventional energy systems promises sizable economic rewards. Large power plants and oil pipelines capitalize on the economies of scale – they generate electricity and deliver crude oil to customers at a lower cost than many distributed facilities and rail. But these rewards come at a steep price of fundamental safety that is frequently borne by people who have received little or no such economic rewards. It was extremely painful to see Japan get hit by the Fukushima disaster. However, an argument can be made that Japanese society accepted the safety risks associated with nuclear power through the democratic political process, guaranteeing participation of all stakeholder groups. On the other hand, Denmark never had a nuclear power plant within its borders.[81] Moreover, the Danish parliament outright rejected the idea of having one in 1985.[82] Yet the Danes got a taste of the Soviet "Peaceful Atom" as the Chernobyl radioactive cloud rolled through Europe in 1986.[83] Thus, we find the lack of offsetting benefits and acceptance of safety risks to be major energy justice implications of conventional energy technologies.

Although *some* energy accidents have a long reach, *many* accidents affect the adjacent area. And if the faulty infrastructure is located in a populated area, the adjacent communities are affected, too. In the previous chapter, we highlighted that it is rare to find affluent communities near refineries or power plants. Instead, it is poor communities that share their space with accident-prone energy infrastructure. One can argue that people always have a choice where to live and those who choose to live near an energy facility have accepted the risk of an accident. However, in many situations, it is hardly a choice, as the alternative is not having a roof over one's family's head. The disproportional exposure of poor communities to energy accidents is thus another energy justice implication related to the safety of conventional energy technologies.

Vulnerability and nuclear proliferation

When one leaves a house door unlocked or a valuable item in plain sight in a dangerous neighborhood, he exposes his dwelling or belonging to a higher risk of loss. If the house does not contain anything except for the usual household items and the belonging is a piece of jewelry, the risk associated with the potential loss stays with the forgetful person. It is a different story if the house has a switch that can turn off the power to the entire neighborhood and the belonging is an explosive device capable of significant damage. In those cases, the risk goes far beyond the economic situation of the scatterbrained individual. He put others in danger by making a dangerous object more vulnerable to an incursion and acquisition. Conventional energy systems are such dangerous objects and are inherently susceptible to manmade disruptions.

Vulnerability concerns related to nuclear proliferation are especially critical due to the harm that a weapon powered by or containing nuclear material can cause. Actual incidents of nuclear theft and procurement of fissile material are frightening, to say the least. There is no shortage of terrorist groups eager to acquire the nuclear waste or fissile material needed to make a crude nuclear device or a dirty bomb.[84] Former United Nations Secretary General Kofi

Annan put it best when he stated that "nuclear terrorism is still often treated as science fiction – I wish it were. But unfortunately we live in a world of excess hazardous materials and abundant technological know-how, in which some terrorists clearly state their intention to inflict catastrophic casualties."[85] The risks are not confined to the reactor site. All stages of the nuclear fuel cycle are vulnerable, including:

- stealing or otherwise acquiring fissile material at uranium mines;
- attacking a nuclear power reactor directly;
- assaulting spent fuel storage facilities;
- infiltrating plutonium stores or processing facilities;
- intercepting nuclear materials in transit; and
- creating a dirty bomb from radioactive tailings.

This once provoked the Nobel Prize-winning nuclear physicist Hannes Alfven to remark, "Atoms for peace and atoms for war are Siamese twins."[86] The four countries with the largest reprocessing fleets – Belgium, France, Germany, and UK – declared more than 190 tons of separated plutonium in 2007, mostly stored in plutonium dioxide powder at above-ground sites and fuel manufacturing complexes – enough for 20,900 nuclear weapons.

The vast majority of the security incidents, interestingly, regard not the theft of material but actual bombings, armed incursions, acts of terrorism, and sabotage. According to the database of the International Policy Institute for Counter Terrorism, 167 terrorist incidents involving a nuclear target occurred during the period from 1970 to 1999. In Europe, nuclear installations were subjected to 10 terrorist attacks between 1966 and 1977. In the United States, there were 240 bombing threats and 14 actual and attempted bombings of nuclear facilities from 1969 to 1975. Socioeconomic and political instability leads to increases in attempts to use radioactive material as a threat. For example, in Russia, 50 instances of nuclear blackmail were carried out between 1995 and 1997.[87]

Security systems are designed and operated by people, and people make mistakes. Any sufficiently large industry will have its share of poor hiring, flawed background checks, sabotage, fraud, falsification, and laziness. However, the nuclear industry does not serve hamburgers, sell cars, or manufacture furniture – it works with unique radiological and fissile materials capable of causing catastrophic harm if mishandled. And this critical fact sometimes is missing from the nuclear industry's modus operandi. For example, alcoholism and drug abuse are prevalent at Russia's nuclear power plants and reprocessing facilities. One Russian sociologist even commented that "a nuclear power plant does not fight alcoholism, it propagates it. Alcoholics are advantageous for nuclear power plants, they are modest and undemanding."[88] The Ministry of Atomic Energy of the Russian Federation went so far as to say that the nation's nuclear industry had "a total lack of a culture of security."[89]

Much of our existing energy infrastructure is perpetually vulnerable to deliberate and accidental disruptions. The average power plant, for example, delivers its electricity a distance of 220 miles.[90] A comprehensive, three-year Department of Defense study concluded that relying on centralized plants to transmit and distribute electric power created unavoidable (and costly) vulnerabilities. The study noted that T&D systems constituted "brittle infrastructure" that could be easily disrupted, curtailed, or attacked. One of the authors, physicist Amory Lovins, who is currently helping the U.S. military streamline energy-intensive sectors, has long argued that if you build an inefficient, centralized electricity system, major failures, whether by accident or malice, become inevitable by design. In Britain during the coal miner strikes of 1976, a leader of the power engineers famously told Lovins that "the miners brought the country to its knees in eight weeks, but we could do it in eight minutes."[91] This is because the massive, complex, and interconnected infrastructure needed to distribute power from a centralized generation source is brittle and subject to cascading failures easily induced by severe weather, human error, sabotage, or even the interference of small animals. "Electrical grids," notes Lovins, "distribute a form of energy that cannot readily be stored in bulk, supplied by hundreds of large and precise machines rotating in exact synchrony across a continent, and strung together by a frail network of aerial arteries that can be severed by a rifleman or disconnected by a few strikers."[92]

The conclusion of Lovins complements a similar study undertaken by the IEA, which noted that centralized energy facilities create tempting targets for terrorists because they would need to attack only a few, poorly guarded facilities to cause large, catastrophic power outages.[93] Thomas Homer-Dixon, chair of Peace and Conflict Studies at the University of Toronto, cautions that it would take merely a few motivated people with minivans, a limited number of mortars, and a few dozen standard balloons to strafe substations, disrupt transmission lines, and cause a "cascade of power failures across the country," costing billions of dollars in direct and indirect damage.[94] A deliberate, aggressive, well-coordinated assault on the electric power grid could devastate the electricity sector and leave critical sectors of the economy without reliable sources of energy for a long time. Paul Gilman, former executive assistant to the Secretary of Energy, has argued that the time needed to replace affected infrastructure would be "on the order of Iraq," not "on the order of a lineman putting things up a pole."[95]

As the discussion above indicates, conventional energy systems are *inherently* vulnerable to manmade disruptions. The examples given in this chapter show that attacks on energy installation are *not* highly unlikely. In addition, the scope and scale of potential damage, especially in the case of nuclear proliferation, may reach catastrophic proportions. Thus, according to probabilistic risk theory, which defines risk as probability multiplied by severity, risks associated with human infractions on energy infrastructure should be placed in the "high" category.[96] However, it is not simply the high-risk nature of conventional energy facilities that poses energy justice concerns. After all, people deal with high-risk situations all the time. A police officer knowingly takes on the risks of patrolling the streets. A patient going for a planned surgery is extensively briefed on a whole host of risks associated with the medical procedure. However, when a company makes the decision to build a 1 GW nuclear power plant or a transnational natural gas line, it shifts the risks associated with this infrastructure onto society. The size of the damage to the company whose facility is compromised may be negligible to that of the community that surrounds the facility. Financial loss due to interruption of service may sting, but lost lives, destroyed homes, and days without power, water, and heat will devastate a community and region. And with the threat of nuclear proliferation, a small but determined terrorist group can hold entire countries hostage. Thus, such a risk shift is an inherent justice implication of conventional energy systems.

Path dependence and carbon lock-in

Discussion of the technological dimension of energy justice would not be complete without examining an aspect of modern energy systems that may dwarf what we have thus far discussed in this chapter – path dependence or carbon-lock in.

Social sciences define path dependence as a phenomenon in which players continue to operate under conditions that are no longer valid. The rationale behind path dependence is very simple: initial conditions influence and sometimes determine outcomes.[97] Douglas Puffert, from the University of Warwick, defines path dependence as follows:

> Path dependence is the dependence of economic outcomes on the path of previous outcomes, rather than simply on current conditions. In a path dependent process, "history matters" – it has an enduring influence. Choices made on the basis of transitory conditions can persist long after those conditions change. Thus, explanations of the outcomes of path-dependent processes require looking at history, rather than simply at current conditions of technology, preferences, and other factors that determine outcomes.[98]

Path dependence often results in the creation of positive feedback loops among several parts of the system.[99] Whether it is a layout of computer keyboards[100] or the national economy, such a system resists any change,[101] making path dependence a "self-perpetuating process."[102] Path-dependent societies "embody belief systems and institutions that fail to confront and solve new problems of societal complexity."[103]

Conventional energy systems pose an easy target for path dependence.[104] Large facilities and extensive infrastructure require large-scale investments. Large investments require long periods during which the investment is recouped, thereby ensuring the longevity of path dependence.[105] The dominance of fossil fuel has led many authors to the adoption of the term "carbon lock-in" in relation to path dependence in the energy sector.[106] According to Gregory C. Unruh, "Carbon lock-in arises through a combination of systematic forces that perpetuate fossil fuel-based infrastructures in spite of their known environmental externalities and the apparent existence of cost-neutral, or even cost-effective, remedies."[107]

A broad range of analysts and policymakers, including the world's most prominent energy experts, has recognized the critical importance of avoiding carbon lock-in for the world's future. The IEA cautioned policy makers that if they keep relying on conventional technological solutions, the global community would find itself locked into an "insecure, inefficient and high-carbon energy system."[108] "As each year passes without clear signals to drive investment in clean energy, the 'lock-in' of high carbon infrastructure is making it harder and more expensive to meet our energy security and climate goals," stated Fatih Birol, IEA Chief Economist.[109] As noted, the IEA came up with the 2017 carbon lock-in deadline, if sweeping energy efficiency measures are not implemented.[110]

Why is carbon lock-in unjust? Because it would affect *everyone* on the planet, making catastrophic effects of global climate change irreversible. It would affect a financial institution that hoped to bring a reasonable return on investment in a new massive coal-fired power plant. It would affect the people who worked for this institution. It would affect the owners and workers of the aluminum smelter that the plant powered. It would affect people who lived in the vicinity of the plant and people who lived on the other side of the mountain ridge the plant was located nearby. It would affect the citizens of the country that opened its doors to the new shiny concrete, steel, and glass-engineering marvel. It would also affect citizens of bordering countries and citizens of countries halfway across the globe. It would affect proponents of renewable energy and its opponents. It would affect the authors of this book and their readers, as well as their children, and their children's children.

Carbon lock-in deprives all the aforementioned people of choices and changes once the point of no return is reached. It would likely not matter if the global community decided to make renewable energy its primary energy technology once the Greenlandic ice sheet had melted and massive quantities of methane had been released in the Canadian and Russian tundra. It also invalidates the choices and sacrifices that many have made to move toward a low-carbon future. Because of the IEA's 2017/2022 cut-off point, carbon lock-in is hardly a condition of the future. Unless we divorce ourselves from coal-fired power plants and oil pipelines, we will ensure that we are battered by the catastrophic impacts of global climate change.

Conclusion

The technologies that convert energy fuels into usable services play a significant role in distribution of benefits and externalities. Some technologies encourage or allow one to get "what she paid for" and others shift the burdens onto others. Technological deficiencies of centralized, fossil fuel-based systems clash with both the prohibitive and affirmative principles of energy justice. A neighborhood located close to an energy facility has a higher risk of being affected by an accident (the prohibitive principle). Energy efficiency improvements to the grid leave people who do not have access to electricity in the same disadvantaged situation (the affirmative principle).

To repeat again, we are not suggesting the conventional systems have absolutely no place in the energy future. However, it makes little sense to copy a system whose foundation is hindered by several crumbling blocks. Thus, we hope the following observations that identify such faulty pieces are given attention when the energy system is overhauled:

- Energy systems deployed in the developing world are usually less efficient than those of developed countries.
- Energy infrastructure usually lacks a basic level of reliability and safety and is vulnerable to manmade interruptions.
- Developed countries employ generally more efficient systems, but the advantage is often caused by the difference in fuel mix.
- Although energy infrastructure deployed in the developed world performs in a more reliable and safe manner, it often fails to meet the requirements of a more advanced economy because service interruptions and accidents lead to higher economic costs.
- Energy facilities in the developed world do not display the same degree of vulnerability than in developing countries, but the level of severity of potential consequences puts them in a higher-risk category.
- Duplication of conventional systems employed in the developed world in developing countries will not ensure growth in energy demand in a more efficient, reliable, safe, and less vulnerable manner.
- Efficiency improvements in conventional energy systems alone will not necessarily improve equity energy access.
- Energy efficiency is capable of postponing carbon lock-in, the ultimate instance of energy justice, but not preventing it.
- Conventional energy systems interfere with and sometimes preclude proliferation of distributed technologies that are capable of improving reliability.
- Conventional energy systems do not distribute the cost of reliability equally, and make it financially prohibitive for low-income groups to employ self-help, such as diesel generators.
- The cost of accidents is often borne by people who did not cause them or assume the safety risk.
- Because poor communities are frequently located close to energy infrastructure, they are disproportionally exposed to energy accidents.
- Operators of conventional energy facilities shift risks arising from infrastructure vulnerability onto adjacent communities and society at large.
- Centralized, fossil fuel-based energy systems have pushed the world to the brink of carbon lock-in, the worst technological energy justice implication.

Notes

1. "Cyberwar: War in the Fifth Domain," *The Economist,* July 3, 2010, 20–22.
2. For example, currently 40.4 percent of electricity is generated from coal and 26.9 percent from natural gas. U.S. EIA, "Electricity, U.S. Electricity by Fuel, All Sectors," http://www.eia.gov/electricity/.
3. See generally, Bruce Rich, *Foreclosing the Future: Coal, Climate and Public International Finance* (Environmental Defense Fund, 2009).
4. IEA, "World Energy Outlook, Executive Summary 1" (2012) (hereafter, "WEO 2012").
5. BP Energy Outlook, January 2013.
6. IEA, "Worldwide Trends in Energy Use and Efficiency 27," 43 (2008).
7. Ibid., 73.
8. Ibid., global average efficiency of natural gas – based power generation is 40 percent.
9. Ibid.
10. United Nations Development Programme, *Human Development Report 2010* (New York: UNDP, 2010).
11. The Chairman of the Presidential Committee on Power Sector Reforms, Dr. Rilwanu Lukman, indicates that "the average age of all the transformers, generating stations and sub-stations in the country is twenty five years"; the aging infrastructure has been made worse again, he said, by a "poor maintenance culture." Dr. Lukman estimates that $85 billion dollars is needed to overhaul and fix the decrepit power sector, a sum that does not even include related gas infrastructure needs. "$85 Billion Needed for Stable Power; Supply Won't Improve Till December," *Daily Trust/All Africa Global Media*, June 25, 2008 (provided by Comtex Online) http://pro.energycentral.com/professional/news/power/news_article.cfm?id = 10554454.
12. World Bank, "Project Paper Proposed Additional Financing Credit in the Amount of SDR $49.6 Million and a Proposed Additional Financing Grant in the Amount of SDR $10.5 Million to Nepal for the Power Development Project," Report No. 48516-NP (Washington, DC: World Bank Group, 2009).
13. Anthony D'Agostino and Benjamin Sovacool, "Energy Security: Introduction," *Asian Trends Monitoring Bulletin* 1 (2010): 21–26.
14. Damages for these accidents have been updated to 2010 USD.
15. See Stefan Hirschberg, Gerard Spiekerman, and Roberto Dones, *Severe Accidents in the Energy Sector*, PSI Report No. 98–16 (Villigen, Switzerland: Paul Scherrer Institute, 1998); Stefan Hirschberg and Andrej Strupczewski, "Comparison of Accident Risks in Different Energy Systems: How Acceptable?" *IAEA Bulletin* 41 (1999): 25–30; Stefan Hirschberg, Peter Burgherr, Gerard Spiekerman, and Roberto Dones, "Severe Accidents in the Energy Sector: Comparative Perspective," *Journal of Hazardous Materials* 111 (2004): 57–65.
16. Note that South Korea only recently "graduated" from the ranks of developing nations. International Monetary Fund, "World Economic Outlook: Growth Resuming, Dangers Remain," IMF Report 180, April 2012.
17. China Labour Bulletin, "Bone and Blood: The Price of Coal in China," in *Sparking a Worldwide Energy Revolution: Social Struggles in the Transition to a Post-Petrol World,* ed. Koyla Abramsky (Oakland, CA: AK Press, 2010), 406–423.

18. Matthew Bunn and Eben Harrell, *Consolidation: Thwarting Nuclear Theft* (Cambridge, MA: Report for Project on Managing the Atom, Belfer Center for Science and International Affairs, Harvard Kennedy School, 2012).

19. Government Accountability Office, "Combating Nuclear Smuggling: Corruption, Maintenance, and Coordination Problems Challenge U.S. Efforts to Provide Radiation Detection Equipment to Other Countries, vol. 7" (2006), 481 cases; Renssalaeer Lee, "Report for Congress, Nuclear Smuggling and International Terrorism: Issues and Options for U.S. Policy CR4" (2002), 426 cases. See also B. K. Sovacool, "Contesting the Future of Nuclear Power and Marina Koren, 'Top Ten Cases of Nuclear Theft Gone Wrong,'" *Smithsonian Magazine*, February 4, 2013.

20. Kishore Kuchibhotla and Matthew McKinzie, *Nuclear Terrorism and Nuclear Accidents in South Asia* (Washington, DC: Stimson Center, 2004), 30.

21. IAEA Incident and Trafficking Database (ITDB), February 21, 2013, http://www-ns.iaea.org/security/itdb.asp.

22. Vijay Sakhuja, "Securing the Nuclear Energy Supply Chain: The Maritime Dimension," paper presented to the Emerging Challenges to Energy Security in the Asia Pacific International Seminar, Center for Security Analysis, Chennai, India, March 16–17, 2009.

23. See B. K. Sovacool and M. A. Brown, "Competing Dimensions of Energy Security: An International Review," *Annual Review of Environment and Resources* 35 (2010): 77–108; as well as B. K. Sovacool, "Defining, Measuring, and Exploring Energy Security," in *The Routledge Handbook of Energy Security*, ed. B. K. Sovacool (London: Routledge, 2010), 1–42.

24. The Abqaiq oil processing plant is the largest facility in the world for removing hydrogen and sulfur dioxide from crude oil.

25. See B. K. Sovacool, *Routledge Handbook of Energy Security*.

26. Jennifer Giroux, *Revisiting the Niger Delta: Energy Infrastructure Threatened* (Washington, DC: US Institute for Peace, 2012).

27. Adam Nossiter and Scott Sayare, "Militants Seize Americans and Other Hostages in Algeria," *The New York Times*, January 16, 2013; "Fate of U.S. Hostages in Algeria," *CNN*, January 18, 2013.

28. See Marilyn A. Brown, Frank Southworth, and Terese K. Stovall, *Towards a Climate-Friendly Built Environment* (Washington, DC: Pew Center on Global Climate Change, 2005); Spencer Abraham, "National Energy Policy Report of the National Energy Policy Development Group," Hearing Before the Subcommittee on Energy and Air Quality of the House Committee on Energy and Commerce (Washington, DC: Government Printing Office, 2005); National Energy Policy Development Group, *Reliable, Affordable, and Environmentally Sound Energy for America's Future* (Washington, DC: Whitehouse Printing Services, 2001).

29. United Nations Development Programme, *Human Development Report 2010*.

30. IEA, "Worldwide Trends in Energy Use and Efficiency."

31. Amory B. Lovins, "Energy Myth Nine: Energy Efficiency Improvements Have Already Reached Their Potential," in *Energy and American Society: Thirteen Myths*, ed. Benjamin K. Sovacool and Marilyn A. Brown (New York: Springer, 2007), 239.

32. Robert Bent, Lloyd Orr, and Randall Baker, eds., *Energy: Science, Policy, and the Pursuit of Sustainability* (Washington, DC: Island Press, 2002).

33. Fereidoon P. Sioshansi, "Can We Have Our Cake and Eat It Too? Energy and Environmental Sustainability," *Electricity Journal* 24, no. 2 (2011): 76–85.
34. Lovins, "Energy Myth Nine."
35. U.S.-Canada Power System Outage Task Force, "Interim Report: Causes of the August 14th Blackout in the United States and Canada," vol. 1 (November 2003).
36. Elisabeth Rosenthal, "Ahead of the Pack on Cleaner Power," *International Herald Tribune,* September 30, 2010, vi.
37. U.S.-Canada Power System Outage Task Force, "Final Report on the August 14, 2003 Blackout in the United States and Canada: Causes and Recommendations," April 2004, https://reports.energy.gov/BlackoutFinal-Web.pdf.
38. Christine Dell'Amore, "Exxon Valdez Anniversary: 20 Years Later, Oil Remains," *National Geographic News,* March 23, 2009; Janet Raloff, "Exxon Valdez Oil Lingers, as Does Its Toxicity," *Science News,* March 24, 2009; Christopher Bergendorff, "20 Years Later: Exxon Valdez Spill Lingers," *Science Central,* March 24, 2009.
39. Benjamin K. Sovacool, "The Costs of Failure: A Preliminary Assessment of Major Energy Accidents, 1907 to 2007," *Energy Policy* 36, no. 5 (2008): 1802–1820.
40. The study defined a "major energy accident" as one that resulted in either death or more than $50,000 of property damage.
41. Amory B. Lovins and L. Hunter Lovins, *Brittle Power: Energy Strategy for National Security* (Andover, MA: Brick House Publishing Company, 1982).
42. Ibid.
43. Ibid., 145.
44. Douglas Pasternak, "A Nuclear Nightmare," *U.S. News and World Report,* September 17, 2001.
45. "A Review of Enhanced Security Requirements at NRC Licensed Facilities," Hearing Before the House Subcommittee on Oversight and Investigations, 108th Congress, Statement of David N. Orrik, Reactor Security Specialist, Nuclear Regulatory Commission (April 11, 2002).
46. "A Report to the Minister of State for Energy, Department of Energy and Climate Change by the Director of Civil Nuclear Security," The State of Security in the Civil Nuclear Industry and the Effectiveness of Security Regulation April 2002–March 2003 (London: OCNS, 2003).
47. The Organization and Work of the Office for Civil Nuclear Security, OCNS, February 2005.
48. See Swiss Federal Nuclear Safety Inspectorate, *Position of the Swiss Federal Nuclear Safety Inspectorate regarding the Safety of the Swiss Nuclear Power Plants in the Event of an Intentional Aircraft Crash,* HSK-AN-4626 (Würenlingen, March 2003); Swiss Federal Nuclear Safety Inspectorate (HSK), "Protecting Swiss Nuclear Power Plants Against Airplane Crash," memorandum, 2003, 7.
49. Ibid.
50. Sovacool, *Routledge Handbook of Energy Security.*
51. Lovins and Lovins, *Brittle Power.*
52. Sovacool, *Routledge Handbook of Energy Security.*
53. Ibid.

54. The Federal Bureau of Investigation, "North to Alaska Part 4: The Shot That Pierced the Trans-Alaska Pipeline," November 23, 2013, http://www.fbi.gov/news/stories/2012/november/north-to-alaska-part-4/north-to-alaska-part-4-the-shot-that-pierced-the-alaska-pipeline.

55. Ibid.

56. Ibid.

57. The International Energy Agency, *World Energy Outlook* (Paris: OECD, 2012), 2.

58. Ibid.

59. Ibid.

60. EIA, "Efficient World Scenario: Policy Framework," 1, http://www.worldenergyoutlook.org/media/weowebsite/energymodel/documentation/Methodology_EfficientWorldScenario.pdf.

61. Only 9 percent of Uganda's population has access to electricity. The World Bank, "Data, Access to Electricity (% of Population)," http://data.worldbank.org/indicator/EG.ELC.ACCS.ZS/countries.

62. See generally, United Nations Foundation, *Realizing the Potential of Energy Efficiency: Targets, Policies, and Measures for G8 Countries* (July 2007).

63. See Chapters 3 and 6 for the discussion of the effects of global climate change.

64. "WEO 2012," 3.

65. Ibid.

66. Benjamin K. Sovacool, *The Dirty Energy Dilemma: What's Blocking Clean Power in the United States* (Westport, CT: Praeger, 2008).

67. Quoted in ibid.

68. National Research Council, *Making the Nation Safer: The Role of Science and Technology in Countering Terrorism* (Washington, DC: National Academies Press, 2002), 178.

69. IHS CERA, "Future Market for Canadian Oil Sands, Special Report," 5–8.

70. Javier Blas, "Libya's Impact on Oil," *Financial Times*, February 23, 2011, http://www.ft.com/cms/s/0/08b1207c-3f29–11e0–8e48–00144feabdc0.html#axzz2Qdq0iWfV.

71. U.S. EIA, "Countries, Libya," http://www.eia.gov/countries/cab.cfm?fips=LY.

72. IEA, "Oil Market Report," *Annual Statistical Supplement* 4 (August 2011).

73. IPCC, "Renewable Energy Sources and Climate Change Mitigation Special Report of the Intergovernmental Panel on Climate Change" 612 (2011).

74. See Fred Bosselman et al., *Energy, Economics and the Environment: Cases and Materials*, 3rd ed. (2010), 526 –528.

75. See Chapter 6 for more on this point.

76. The architects of this database define a "severe accident" as one that involves one of the following: at least 5 fatalities, at least 10 injuries, 200 evacuees, 10,000 tons of hydrocarbons released, more than 25 square kilometers of cleanup, or more than $5 million in economic losses.

77. The study defined a "major energy accident" as one that resulted in either death or more than $50,000 of property damage.

78. David Lochbaum and Union of Concerned Scientists, "U.S. Nuclear Plants in the 21st Century: The Risk of a Lifetime" (2004), 5.

79. See Eric S. Beckjord et al., *The Future of Nuclear Power: An Interdisciplinary MIT Study* (Cambridge, MA: MIT Press, 2003), 22, 48.

80. U.S. Government Accountability Office, *Additional Guidance and Oversight of Mines' Emergency Response Plans Would Improve the Safety of Underground Coal Miners,* GAO-08–424 (April 2008).

81. World Nuclear Association, "Nuclear Energy in Denmark," January 2013, http://www.world-nuclear.org/info/Country-Profiles/Countries-A-F/Denmark/#.UW_rJrVvOZd.

82. Ibid.

83. See generally Ian Fairlie and David Sumner, *The Other Report on Chernobyl (Torch)* (European Parliament, April 2006), http://www.chernobylreport.org/torch.pdf.

84. Steve Bowman, "Weapons of Mass Destruction: The Terrorist Threat," *CRS Report for Congress*, RL31332 (Washington, DC: Congressional Research Service, 2002).

85. Kofi Annan, address to the General Assembly of the United Nations, March 10, 2005.

86. Quoted in Alexander Shlyakhter, Klaus Stadie, and Richard Wilson, *Constraints Limiting the Expansion of Nuclear Energy* (New York: United States Global Strategy Council, 1995).

87. Greenpeace, *Media Briefing: Nuclear Power and Terrorism* (London: Greenpeace United Kingdom, 2006).

88. Mohammad Saleem Zafar, *Vulnerability of Research Reactors to Attack* (Washington, DC: Stimson Center, 2008), 19–20.

89. Ibid.

90. Sovacool, 2008.

91. Lovins and Lovins, *Brittle Power*.

92. Ibid.

93. International Energy Agency, *Energy Security in a Dangerous World* (Paris: International Energy Agency, 2004).

94. Thomas Homer-Dixon, "The Rise of Complex Terrorism," *Foreign Policy* (January/February 2002): 34–41.

95. Quoted in Sovacool, *Dirty Energy Dilemma*.

96. U.S. Nuclear Regulatory Commission, *An Approach for Using Probabilistic Risk Assessment in Risk-Informed Decisions on Plant-Specific Changes to the Licensing Basis* (Washington, DC: NRC, 2002).

97. A. Goldthau and B. K. Sovacool, "The Uniqueness of the Energy Security, Justice, and Governance Problem," *Energy Policy* 41 (2012): 232–240.

98. Douglas Puffert, "Path Dependence," *EH.net*, February 1, 2010, http://eh.net/encyclopedia/article/puffert.path.dependence.

99. Goldthau and Sovacool, "The Uniqueness of the Energy Security."

100. See generally Paul A. David, *Path Dependence and the Quest for Historical Economics: One More Chorus of the Ballad of QWERTY,* University of Oxford Discussion Papers in Economics and Social History No. 20 (November 1997), http://www.nuff.ox.ac.uk/economics/history/paper20/david3.pdf.

101. Goldthau and Sovacool, "The Uniqueness of the Energy Security."

102. Douglas C. North, "Economic Performance Through Time," prize lecture, December 9, 1993, http://www.nobelprize.org/nobel_prizes/economics/laureates/1993/north-lecture.html.

103. Ibid.

104. Goldthau and Sovacool, "The Uniqueness of the Energy Security."
105. See, e.g., Roman Sidortsov, "Creating Arctic Carbon Lock-in: Case Study of New Oil Development in the South Kara Sea," *Carbon and Climate Review* 6, no. 1 (2012): 8–9.
106. See ibid.; Brown et al., "Carbon Lock-In: Barriers To Deploying Climate Change Mitigation Technologies" (Oak Ridge National Laboratory, January 2008), ix–xiii, http://www.ornl.gov/sci/eere/PDFs/ORNLTM-2007–124_rev200801.pdf.
107. Gregory C. Unruh, "Understanding Carbon Lock-in," *Science Direct, Energy Policy* 28, no. 2: 817.
108. IEA Press Release, November 2011, London, http://www.worldenergyoutlook.org/media/weowebsite/2011/pressrelease.pdf.
109. Ibid.
110. "WEO 2012," 2.

8 Toward a more just and secure energy future

The fault, dear Brutus, is not in our stars, but in ourselves, that we are underlings.
– Cassius from William Shakespeare's *Julius Caesar,* Act 1, Scene 2

Introduction

We were fully aware that we would deviate from the conventional conversation about energy when we decided to write this book. Our goal was not to make an economic or geopolitical case for or against a certain energy technology. We did not aim to persuade the reader that polar bears may not exist in the near future unless the world changes the way it uses its energy. We chose not to put all the blame for global economic, social, and environmental problems on oil companies and embark on a moral crusade to eradicate fossil fuels from our lives. We premised our approach on the lack of a political and normative agenda when we agreed to "check" our personal views at the door.

We tried to make people the center of our book. *All* sorts of people: rich and poor, young and old, from Ukraine and Uganda to the United Arab Emirates and the United Kingdom. Despite differences in social, economic, and political status, all people are entitled to a minimum of basic goods. These goods include the rights to life, liberty, freedom of movement, security of person, and the capacity to control one's environment. Because energy is so fundamental to all economic activities today, it has become a necessary condition for realizing many of these basic goods. And by virtue of being an instrumental good and material prerequisite, energy plays a major role in the distribution of these goods among people.

After establishing this baseline, other conventions started to crumble. We no longer accepted as bright a line between developed and developing countries, as energy-related inequality can be as prevalent in the United States as in Azerbaijan. The assumed technological superiority of the developed world needed to be qualified, as did the link between energy use and economic growth. We departed from the classic definition of security of energy supply and the even less accepted definition of security of demand. Instead, we became more interested in how much and what kind of energy enables a person to live a truly human life without interfering with or diminishing the ability of others to do the same.

We acknowledged the benefits that energy services bring to people but we chose to focus on the instances where energy is (at least part of) the problem.

The rationale behind this approach was simple – one cannot trace the outline for a model of a just energy system without identifying instances of energy injustice. As this book has catalogued, plenty of these exist in the world. In fact, there are so many of them that we were overwhelmed at first by the sheer numbers and complexity. However, the five central dimensions of energy injustice highlight the profound effect of conventional energy systems on economic, sociopolitical, and environmental wellbeing, as well as the geographic distribution of benefits, burdens, and risks.

In this concluding chapter, we present the book's four key conclusions: (1) there is an often neglected ethical and political dimension to the global energy system, which means that energy problems cannot be addressed by technology alone; (2) we are each complicit in perpetuating many of the energy injustices mentioned throughout this book; (3) the presence of these injustices radically transforms how we should deliberate about energy policy; and (4) an energy justice "checklist" may provide a blueprint for achieving a better energy future. As Cassius put it pithily, the fault lies with us, but so do many of the solutions.

Conclusion 1: Technology cannot solve our energy problems

Perhaps the simplest, and least controversial, conclusion is that technical solutions cannot entirely resolve our collective energy problems. This is partly because energy technologies need to be interpreted not as neutral devices that produce, transport, or convert energy, but as forms of congealed culture where the social interests of those designing the technology get built into the system. The global energy system thus redistributes social power and entrenches established practices that have grave consequences for society, including the emission of greenhouse gases and toxic pollutants. Put another way, the manner in which energy technologies redistribute social, political, and economic power is as important as how such technologies generate electricity or transport liquid fuel. A key implication here is that the global energy system not only satisfies our wants, desires, and needs; it can also influence our behavior. Economist E. J. Mishan captured the Janus face of technology well when he remarked "while new technology is unrolling the carpet of increased choice before us by the foot, it is often simultaneously rolling it up behind us by the yard."[1]

The implication that energy systems are simultaneously technological as well as *political* and *ethical* reminds us that more efficient energy devices and improved economic signals are necessary but not sufficient conditions to achieve energy justice. Utility managers, systems operators, business leaders, and ordinary consumers do not function like automatons that rationally calculate price signals and change their behavior to optimize benefits and minimize costs. Instead, they are embroiled in a complicated social and cultural environment that is shaped by and helps to shape technological change, behaviors, values, attitudes, and interests.

Up until this point many experts have primarily focused on the "energy problem" as a purely technical one, and policymakers continue to rely extensively on technology to provide solutions to what are, in essence, deeper social, political, and cultural problems. The intellectual program of energy policymaking thus seems guided by a fundamental but unacknowledged assumption regarding the priority of technical prescriptions to energy troubles. Beyond the "hard" analysis, modeling, and conjecture about the particular characteristics of conventional and alternative clean electricity technologies, however, the important issues are really moral and political as much as they are technological.

Back in the 1970s, the physicist Amory Lovins argued that the true contest between promoting centralized fossil fuel and nuclear generators and decentralized energy efficiency and renewable systems had little to do with the technologies themselves, and much more about the way that policymakers thought about energy. One could conceivably integrate a collection of 1 kW solar panels on the cooling tower of a large nuclear facility. Yet these two approaches were deemed "culturally incompatible" since "each path entails a certain evolution of social values and perceptions that make the other kind of world harder to imagine."[2]

Thus, the choice between just energy technologies and unjust energy technologies may rely on different orders of thinking – much like the difference between chess and checkers. An energy justice path seeks to incorporate responsibility, fairness, and equity as guiding principles, whereas an unjust path disregards these considerations while making compound growth, an increase in resource extraction, and profit-making the top priorities. Consequently, the choice between energy technologies is about more than merely hardware. It represents a battle between new stakeholders and entrenched interests; a contest over how best to manage power systems, industrial utilities, and the energy use of businesses and firms; and a conflict over competing conceptions of modern life, identity, and, yes, justice. The conflict within the global energy system is at once material and immaterial, institutional as well as technological, social and moral as well as scientific.

Conclusion 2: We are each responsible for energy injustice

Secondly, all of us participate in the global energy system, and thus each of us contributes to some – perhaps even many – of the problems and examples of injustice highlighted throughout this book. We each consume varying levels of energy to warm our homes, cook our meals, and to travel to and from work. Yet the decisions we make about which home appliance to purchase, which electricity company to patronize, and which car to buy have very real moral and ethical implications. The heating oil we purchase contributes to particulate matter being inhaled by children living near refineries. The fuelwood we collect and burn exposes women and children to indoor air pollution and contributes to the deforestation of tropical rainforests. The electricity we consume is partially responsible for the impacts of climate change – the tidal inundations and storm surges flooding our low-lying cities – as well as the forcible relocation of indigenous communities near a coal mine or hydroelectric dam. The automobile we covet contributes to the acid rain bleaching streams, forests, and coral reefs, or to human rights abuses in oil-producing countries.

Because of our own involvement in the energy system, and our collective responsibility for its problems, energy problems inescapably have a technical element, involving technologies and infrastructure, and a moral element, involving how we value other human beings (now and in the future) and the natural environment. This may be part of the reason why the ethicist Peter Singer has called climate change "the greatest *moral* challenge our species have ever faced," and why philosopher Stephen M. Gardiner has called climate change a perfect "*moral* storm."[3]

At present, however, our moral orientation seems unequal to the task of accommodating energy and climate change problems. Interestingly, one recent study from psychologists and environmental scientists at the University of Oregon concluded that human moral systems are not well attuned to address deteriorating energy security and climate change.[4] They noted that cognitively, the concept of climate change is abstract and complex, with non-linear and hard to predict outcomes, and moreover, that as an unintended side-effect of human activity, it is difficult to "blame" anyone for it. Climate change also runs counter to our guilt bias; that is, humans don't like to feel guilty, and will derogate evidence of their own role in causing a problem. The implication of the study's findings is that individuals will work hard to avoid the very feelings of responsibility for climate change that are needed to combat it.

Concerns over individual denial or avoidance of the problems associated with energy consumption are only part of the problem. Our book seriously questions whether many energy companies – and the governments that back them – will strive for energy justice on their own. On the contrary, many of the examples of energy injustice we have described in this book occurred in the face of concerted and vigorous opposition. The countless examples of corruption, social exclusion, human rights abuses, and conflict run counter to initiatives promoting better transparency, accountability, and stakeholder involvement throughout parts of the energy sector. These latter efforts include the creation of independent inspection panels at the World Bank and other multilateral financial institutions, the Extractive Industries Transparency Initiative (EITI), the rise of corporate social responsibility and social entrepreneurship, the norm (recently recognized as a duty under international law by the International Court of Justice[5]) of conducting adequate environmental and social impact assessments, the notion of free prior informed consent (FPIC), and enhanced awareness concerning human rights generally.

All of which brings us to a troubling thought: neither individuals, nor companies and governments, seem able to provide energy justice on their own. Energy projects operating in the "free market" without proper public oversight can leave more destruction and degradation in their wake than lasting economic development and human dignity.

Conclusion 3: We need energy policymaking directed by justice principles

Our third conclusion is that we need to match more ethical, individual, "bottom-up" actions with comprehensive "top-down" changes in energy policymaking. Recognizing justice concerns can radically transform how planners and politicians deliberate and make energy policy.

The incorporation of considerations of justice into energy policymaking will alter how we view entire energy systems, with concerns such as equity and equality of distribution becoming more predominant, while other concerns, such as profit maximization, will recede in importance. Energy consultant Ronald Binz and his colleagues capture some of this thinking in their recent proposal for "risk-aware" regulation of energy.[6] Binz et al. argue that according to traditional cost numbers, most regulators prioritize long-term energy investments according to the left diagram in Figure 8.1 – that is, they will decide to invest in energy efficiency, some renewable forms of energy, natural gas, and nuclear power. However, the authors discovered that when they factored in just a few non-traditional factors such as unplanned cost increases, social opposition, air and water quality rules, greenhouse gas emissions, and water constraints, the entire picture changed to the right diagram in Figure 1: Nuclear and natural gas become much less desirable, whereas solar and wind power, among others, become significantly more attractive. Binz and his colleagues conclude that, as a result of incorporating a broader array of risks, energy regulators must "reform and reinvent" how we price energy and electricity.

Although we don't necessarily agree with the way that Binz et al. conceptualize "risk," we do believe that what is really needed is for energy regulators to recognize and appreciate the concerns of social justice. The ultimate success of smarter, cleaner, and more reliable energy systems cannot be separated from, and will to a certain extent depend upon, how those systems interact with people and the natural environment. Put another way, energy justice demands that we evolve new business models and regulatory paradigms that promote inclusive and transparent planning processes, diverse resource portfolios, and energy policies that respect the future. The community of energy planners needs to shift from imposing negative externalities on marginalized populations and future generations to endorsing those energy systems that focus on bettering social welfare and minimizing environmental damage. The dominant model of energy policy – business as usual, what we're doing now – can only be endorsed uncritically if one has extremely limited criteria for endorsement.

The prohibitive and affirmative principles of energy justice articulated in Chapter 2 can direct energy policy thinking as we move forward in developing new regulatory paradigms. The prohibitive principle states that *energy systems must be designed and constructed in such a way that they do not unduly interfere with the ability of people to acquire those basic goods to which they are justly entitled.* In other words, the prohibitive principle requires regulators to focus on the burdens or costs that result from the energy system, and how these burdens are distributed across present and future generations. All energy systems have associated indirect costs, of course. Wind farms spoil the landscape for some, and result in the deaths of significant numbers of birds and bats. The production process for photovoltaics generates toxic waste. For this reason the prohibitive principle states that energy systems should not "unduly" impose upon the environment and human health.

What is required is a more nuanced balancing of the benefits and the burdens of a proposed project, one that also examines *who* is affected by the benefits and

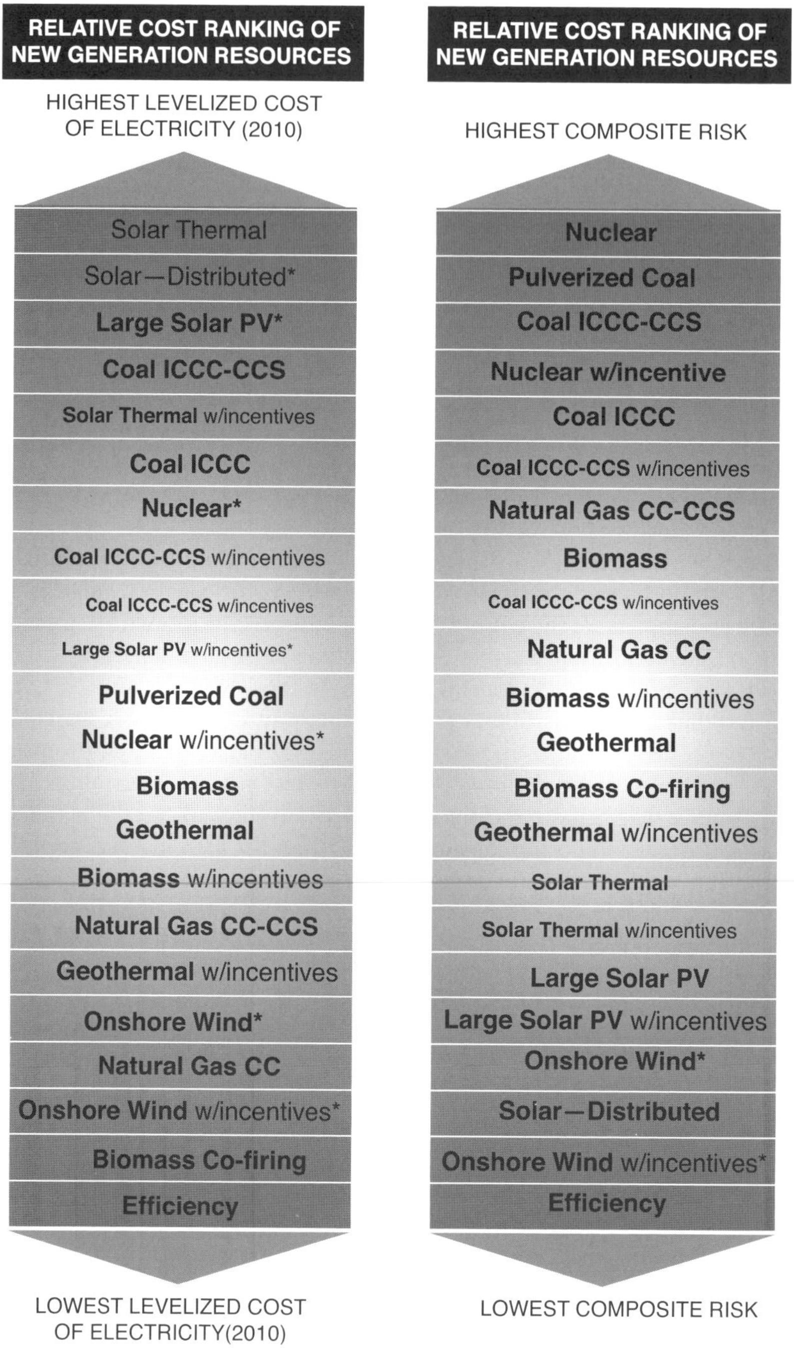

RELATIVE COST RANKING OF NEW GENERATION RESOURCES	RELATIVE COST RANKING OF NEW GENERATION RESOURCES
HIGHEST LEVELIZED COST OF ELECTRICITY (2010)	HIGHEST COMPOSITE RISK
Solar Thermal	Nuclear
Solar—Distributed*	Pulverized Coal
Large Solar PV*	Coal ICCC-CCS
Coal ICCC-CCS	Nuclear w/incentive
Solar Thermal w/incentives	Coal ICCC
Coal ICCC	Coal ICCC-CCS w/incentives
Nuclear*	Natural Gas CC-CCS
Coal ICCC-CCS w/incentives	Biomass
Coal ICCC-CCS w/incentives	Coal ICCC-CCS w/incentives
Large Solar PV w/incentives*	Natural Gas CC
Pulverized Coal	Biomass w/incentives
Nuclear w/incentives*	Geothermal
Biomass	Biomass Co-firing
Geothermal	Geothermal w/incentives
Biomass w/incentives	Solar Thermal
Natural Gas CC-CCS	Solar Thermal w/incentives
Geothermal w/incentives	Large Solar PV
Onshore Wind*	Large Solar PV w/incentives
Natural Gas CC	Onshore Wind*
Onshore Wind w/incentives*	Soiar—Distributed
Biomass Co-firing	Onshore Wind w/incentives*
Efficiency	Efficiency
LOWEST LEVELIZED COST OF ELECTRICITY(2010)	LOWEST COMPOSITE RISK

Figure 8.1 "Traditional" energy planning and "risk-aware" energy planning
Source: Binz et al.

burdens, and how much choice those affected have in the matter. What is "due," after all (in this case, an equitable sharing of burdens and benefits), can only be decided by an open process in which those to whom such equity is due are consulted. The prohibitive principle, in other words, also indirectly requires of policymakers and regulators that they comply with the principles of public participation and free prior informed consent. Obviously this is more difficult when the consent required involves people not yet born. This circumstance, however, should only serve to make those responsible for decisions more cautious in their deliberations. No harm would come from a more practical and genuine engagement with the precautionary principle, especially in the energy sector.

The affirmative principle of energy justice states that *if any of the basic goods to which people are justly entitled can only be secured by means of energy services, then in that case there is also a derivative entitlement to the energy services.* This principle focuses on how the benefits of energy systems are distributed across societies. It requires regulators to acknowledge that gross inequalities of access to energy exist in the modern world, within and between countries, and that these inequalities have important ramifications for the economic, social and political opportunities that are available to people. Social justice cannot ignore the material prerequisites that must be satisfied if there is to be a meaningful distribution of civil and political rights. In many cases, energy services of some kind constitute a part of the material prerequisites for the possibility of participating in society. Given the importance of energy within the modern world, it is part of the responsibility of regulators to ensure that the energy system works to forward the goals of justice and equality, rather than exacerbate already existing inequalities.

Conclusion 4: An energy justice "checklist" can guide energy decisions

Our fourth conclusion is that we need new ways of approaching the world's energy problems that markets and institutions fail to tackle. So instead of merely rehashing our findings from the preceding chapters, and bemoaning the injustices of the global energy system, this final section of the chapter presents a tool to those who are in a position to make energy decisions, a tool for assessing such decisions from the standpoint of justice, equality, and security. From what we have already said, it should be clear that in some sense we are all energy decision-makers. Many of our readers, who are not energy planners, corporate executives, government regulators, or politicians, do have something to say as voters and members of civil society. Their individual voices will not have the same effect as that of a government agency official or company CEO, but collectively they can make a difference.

Thus, we leave our readers with an energy justice checklist summarized by Table 8.1. The questions below incorporate both the affirmative and prohibitive principles of energy justice we proposed in Chapter 2. The list can be applied to a wide range of items ranging from individual projects to government policies. This checklist is not exhaustive; we hope to expand and elaborate it, and we invite our readers to join us.

Table 8.1 An energy justice checklist

	Question	Example of injustice to be avoided
Temporal	Does a decision to build energy infrastructure account for the physical risks posed by climate change?	A state utility chooses to build a coal-fired power plant near the sea coast. Not only do the emissions from the plant provide a small but real contribution to climate change, but a hurricane and related storm surge destroy the plant. The ratepayers are left without electricity and with the costs of a no longer "used and useful plant."
	Does an energy-related decision account for regulatory risks associated with climate change?	A national oil company invests in the development of an offshore oil field with the expectation of price of oil above $100 per barrel. Subsequently, nations reach a climate agreement with universal controls. The market reacts accordingly, sending the oil price below $100. The NOC's oil lacks a market and the taxpayers are left to absorb the cost of the poor investment.
	Does a country's energy policy interfere with its responsibility to aid victims of climate change?	To accommodate a growing industrial sector a country invests heavily in coal-fired power plants. The emissions from these plants provide a substantial contribution to climate change. At the same time, highly populated coastal areas suffer from flooding due to sea level rise. The country spends significant budget funds assisting people who have lost their homes.
	Will construction of an energy facility exacerbate the effect of climate change?	A newly constructed hydroelectric facility generates renewable power. However, the river no longer reaches a region experiencing water shortages due to climate change.
	Does a plan to build a nuclear waste storage facility provide for long-term risk mitigation measures?	Twenty years after its construction, the facility suffers structural damage. Because the facility is not insurable, taxpayers are left to pay for repairs and cleanup costs.
	Does a decision to subsidize a certain energy industry add to the cost of externalities?	The tax code allows for deduction of intangible costs incurred during exploratory drilling. The deduction further distorts the price of gasoline.
	Does the cost of owning and operating a motor vehicle fully reflect the external cost caused by air pollution?	A driver commuting to work and driving a car with the same gas mileage as a carpooling driver is responsible roughly for the same amount of pollution. Yet the latter does not get rewarded for "preventing" her passengers from driving and, thus, causing more emissions.

Table 8.1 (Continued)

	How does energy affect the distribution of wealth in a country?	A significant portion of the country's population lives below the poverty line while a few citizens connected to the energy industry are among the world's richest people.
	Does the mix of electric generation facilities match the actual demand?	Ratepayers in a non-restructured electricity market are usually left paying for the excessive capacity.
	Does switching to a more sustainable energy mix disproportionately affect poor members of the society?	On one hand, fuel assistance programs help poor communities to stay warm during cold months. On the other hand, these programs perpetuate the elasticity of demand for fossil fuels by discouraging the switch to cleaner heating units. Additionally, environmental externalities associated with combustion of fossil fuels disproportionally affect poor communities.
Economic	To what extent does the construction of an electrical generation facility and/or automobile fleet expansion account for volatility of oil and gas prices?	A utility switches its generation base to new natural gas-fired facilities. Natural gas prices, as they periodically do, rise significantly. Rate payers' electricity bills skyrocket.
		A town police department updates its fleet with large SUVs citing their operational superiority and relying on, at the time, low gasoline prices. Gasoline prices go up and the town budget is diverted from the retrofit of the elementary school to keeping police vehicles patrolling the streets.
	To what extent does the construction of an electrical generation facility account for a long-term increase in electricity prices?	An utility invests in the construction of a natural gas-powered plant. The investment takes 20 years to repay. Due to the overall increase in the fuel price, the electricity produced at the plant is significantly more expensive than that of a large wind farm. Because the utility operates in a non-restructured market, the ratepayers are stuck with high electricity bills.
	What is the full price of military presence in oil producing regions?	The oil industry is subsidized through the defense budget and with the lives and health of national troops.

(Continued)

Table 8.1 (Continued)

	Question	Example of injustice to be avoided
Sociopolitical	How are energy revenues allocated?	All oil and gas revenues go directly to the "general fund." Further allocation of funds is unknown but school teachers have not been paid for three months.
	How transparent is the allocation of energy revenues?	A country has the following budgetary funds that receive revenue from oil exports: the strategic fund, the stabilization fund, the development fund, the education fund, and the general fund. However, the information about the amount in each fund is not publically available. The country's security service just updated its vehicle fleet. School teachers have not been paid for six months.
	To what extent does an energy company rely on the support of the central government known for its questionable governance methods?	The central government uses pipeline construction as an excuse to displace and thus weaken an ethnic community that opposes government's policies.
	What is the difference in the socio-economic status of government elites and an average citizen of the country?	The dominance of the current regime will likely continue as the bureaucrats will be reluctant to worsen their socio-economic status.
	What is the percentage of energy revenues spent on acquiring weapons?	School teachers have not been paid for a year.
	Does the legal and regulatory regime of a country whose economy is dependent on export of oil and gas provide for meaningful participation of its citizens in the decision-making process related to energy matters?	Members of an indigenous community wake up to the sound of bulldozers clearing land for drilling pad.
	How transparent and attentive are the government and involved energy companies to human rights abuses in the energy sector?	The official government TV channel broadcasts an interview with "a member" of the village who is praising the government for bringing a transnational pipeline in the region. Away from the camera, a platoon of soldiers rounds up a family whose house stands in the way of the pipeline.
	Does the issue of the control over energy resources play any role in military conflict?	People sacrifice their lives for the resource that already comes with a hefty environmental and social price.

Table 8.1 (Continued)

Is the facility operator or the industry adequately capitalized to absorb all possible costs of an accident?	The costs of damage from natural gas explosions caused by an attack on an LNG facility exceed the funds that the operator and its insurers are able to provide. The taxpayers are left with the bill.
Does a proposed technology lock out low-carbon solutions?	An independent power producer builds a coal-fired base load power plant instead of a few natural gas load following units capable of compensating for intermittency of non-dispatchable renewable facilities.
Does a fossil fuel-centric project require a multibillion dollar investment that would take decades to repay?	An oil company chooses to invest in exploration and development of new, remote, large, and marginally profitable oil fields. The alternative would have been to further diversify the company's portfolio from hydrocarbon production and invest in enhanced recovery from its mature assets.

Source: Authors.

Conclusion

In aggregate, our four conclusions suggest that to achieve a more just and secure energy system, and energy future, we need to reject the idea that better technology can address all of our energy problems. We need to accept our own complicity for the injustices inherent in the global energy system, and then embark upon making better "bottom-up" personal energy decisions in tandem with improved "top-down" "justice-aware" energy regulation. Indeed, accepting these four conclusions will at least ensure that "justice" remains front and center in discussions about energy technology, production and use. It could also meaningfully ensure that we begin the transition from an energy-hungry world to an energy-just one.

And for those that balk at having to change their behavior, or at having to consider the far-reaching implications of their energy decisions, we recall the words of Jürgen Habermas, who wrote almost six decades ago that "in the process of enlightenment, there can be *only* participants." We need to recognize that we are each energy agents as much as we are energy consumers. Or, as Rachel Carson put it in 1962, "the human race is challenged more than ever to demonstrate our mastery – not over nature, but of ourselves."[7]

Notes

1. E. J. Mishan, *The Costs of Economic Growth* (New York: Praeger, 1967).
2. A. B. Lovins, "A Target Critics Can't Seem to Get in Their Sights," in *The Energy Controversy: Soft Path Questions and Answers,* ed. H. Nash (San Francisco, CA: Friends of the Earth, 1979), 15–34.
3. Quoted on the back cover of Stephen M. Gardiner, *A Perfect Moral Storm: The Ethical Tragedy of Climate Change* (Oxford: Oxford University Press, 2011).
4. Ezra M. Markowitz and Azim F. Shariff, "Climate Change and Moral Judgment," *Nature Climate Change* 2 (2012): 243–247.
5. See Pulp Mills on the River Uruguay (Arg. v. Uru.), (Judgment of April 20, 2010), http://www.icj-cij.org/docket/files/135/15877.pdf.
6. Ron Binz, Richard Sedano, Denise Furey, and Dan Mullen, *Practicing Risk-Aware Electricity Regulation,* Ceres Research Report, April 2012.
7. Rachel Carson, *Silent Spring* (New York: Houghton Mifflin, 1962), ix.

Index

accidents *see* efficiency, reliability, safety, and vulnerability
affirmative justice principle 3, 46–8
affordability *see* resource depletion and rising energy prices
Arctic oil and gas *see* unconventional resources
Aristotle 30, 31, 35

Barry, Brian 31, 45–6
blackout 2, 17, 41, 74, 98, 169, 208
Brown, Marilyn 39, 40
Burke, Edmund 31, 45

carbon lock-in *see* efficiency, reliability, safety, and vulnerability
Chernobyl accident 123, 173, 185
Cicero 33
Clinton, William 116
coal *see* extracting fossil fuels and uranium; inequality and poverty
community displacement 129, 130, 146–50
conflict *see* energy and military conflict
conventional energy systems: and authoritarianism 26, 27, 121–5; definition of 12
corruption, authoritarianism, and energy conflict 16, 115–40; armed conflict, terrorism, and civil war, 130–2; conventional energy systems and authoritarianism 26, 27, 121–5; energy and human rights abuses, 125–30; energy and military conflict, 130–5; energy

profits, corruption, and politics, 117–20; energy resources and interstate war, 132–3; money rules the world, 117–19; nuclear power and public participation, 121–3; secrecy and democracy, 123–5; social marginalization and political instability, 119–21; the technology of modern warfare, 133–5
cosmopolitan justice *see* energy
crude oil *see* extracting fossil fuels and uranium; inequality and poverty

Deepwater Horizon 170, 177
developed countries: definition of 12, 142–3
developing countries: definition of 12, 142–3
discount rate *see* Stern Report
displacement of populations 129, 130, 146–8
distributive justice *see* energy and distributive justice
Dobson, Andrew 31, 46

efficiency, reliability, safety, and vulnerability 16, 160–96; energy technology in the developed world 168–79; energy technology in the developing world 161–8; from structural problems to path dependence 179–89; vulnerability and nuclear proliferation 185–9
electricity 6–7
energy: definition of 10–12